T0189680

Wireless Networks

Series Editor:
Xuemin (Sherman) Shen
University of Waterloo, Waterloo, Ontario, Canada

More information about this series at http://www.springer.com/series/14180

Rongxing Lu

Privacy-Enhancing Aggregation Techniques for Smart Grid Communications

Rongxing Lu
School of Electrical and Electronic
Engineering
Nanyang Technical University
Singapore, Singapore

ISSN 2366-1186 ISSN 2366-1445 (electronic)
Wireless Networks
ISBN 978-3-319-81393-6 ISBN 978-3-319-32899-7 (eBook)
DOI 10.1007/978-3-319-32899-7

Softcover reprint of the hardcover 1st edition 2016

Printed on acid-free paper

This Springer imprint is published by Springer Nature
The registered company is Springer International Publishing AG Switzerland

Preface

Power system safety has been drawing increasing public attention, and there have been extensive efforts in both industry and academia to mitigate the impacts of power system failures. Recent advances in information and communications technology and smart grid evolution bring promising new ways to facilitate power system security and mitigate system failure rate.

The concept of smart grid has been proposed with expectation to provide a coordinated, efficient, reliable, robust, and secure power generation, transmission, and distribution. Precisely, smart grid is comprised of three main parts: smart infrastructure system, smart communication system, and smart management system. The smart infrastructure system is the physical power infrastructure integrated with intelligent electronic devices underlying the smart grid. It supports not only the two-way flow of electricity but also the integration of renewable power resources. The smart communication system is grounded on the wide-area measurement architecture that collects the real-time status across the widespread power system. The smart management system is the most critical subsystem in smart grid, which provides advanced real-time data analysis, operation management, and control services based on the aforementioned two subsystems. Effective and efficient coordination of the three partitions is expected to achieve promising objectives.

However, without strong security and privacy-preserving mechanisms in place, the smart grid will not only inherit the massive existing vulnerabilities of the legacy power grid but also introduce new potential class of vulnerabilities with the integration of various novel technologies. Many security threats (like worms, malwares, insider attacks, etc.) have been reported for the existing power system till now, which result from both physical space and cyber space.

Particularly, secure and privacy-preserving data aggregation is one of the main challenges in smart grid communications. Smart grid is characterized by the real-time data analysis, monitoring, and control, which means massive measurement data of the system status and the customer's detailed electricity usage will be collected and reported to the system control center. This feature will cause considerable communication burden toward the whole network. As a result, an efficient data aggregation scheme is expected to effectively reduce such communication burden in smart grid communications. In addition, the detailed electricity usage data will inevitably expose the customers' personal privacy once they are leaked to unauthorized parties. Therefore, preservation of customer privacy has also been considered as a critical issue in smart grid communications.

In this monograph, we focus on the secure data aggregation and customer privacy issues in smart grid communications. We first provide an overview of security and privacy issues in smart grid communications, as well as the challenges in addressing these issues, and then introduce several privacy-enhancing aggregation techniques for smart grid communications. Note that these privacy-enhancing aggregation techniques naturally can also be applied in other Internet of Things (IoT) scenarios.

There are total eight chapters in *Privacy-Enhancing Aggregation Techniques for Smart Grid Communications*. These are organized as follows:

Chapter 1 introduces the background of smart grid system and identifies its security and privacy challenges, particularly focusing on the data aggregation techniques in smart grid communications.

Chapter 2 discusses two popular homomorphic public key encryption (PKE) techniques, i.e., Paillier PKE and Boneh-Goh-Nissim (BGN) PKE, which serve as the preliminary for building most privacy-enhancing aggregation techniques in the rest of the chapters of this monograph. In addition, the Java source codes of the two homomorphic PKEs are also provided for the interested readers to better understand and implement them.

Chapter 3 presents the first privacy-enhancing aggregation technique in this monograph, i.e., a privacy-preserving multidimensional data aggregation (PPMDA) scheme, which utilizes the unique data characteristics, i.e., nearly real-time data collection and small-size individual data in smart grid, to provide a much efficient data aggregation for smart grid communications.

Chapter 4 proposes another flexible and fine-grained privacy-preserving subset data aggregation (PPSDA) scheme for secure smart grid communications, and the Java source code is also provided at the end of the chapter.

Chapter 5 introduces a multifunctional data aggregation scheme, named MuDA, for privacy-preserving smart grid communications, achieving privacy-preserving aggregation of multiple functions such as average, variance, one-way ANOVA, etc.

Chapter 6 presents a privacy-preserving data aggregation scheme with fault tolerance, named PDAFT, for smart grid communications. Because PDAFT supports the fault-tolerant feature, even when some user failure or server malfunction occurs, PDAFT can still work well.

Chapter 7 discusses a new secure data aggregation scheme, named DPAFT, for secure smart grid communications, which can not only support fault tolerance but also resist against the differential privacy attacks in smart grid communications.

Chapter 8 proposes another secure data aggregation scheme to achieve privacy preservation and data integrity with differential privacy and fault tolerance, obtaining a good tradeoff of accuracy and security of differential privacy for arbitrary number of malfunctioning smart meters.

Finally, I hope this monograph can provide the interested readers with some valuable insights on the design and deployment of future privacy-preserving data aggregation techniques in IoT scenarios in general and smart grid communications in particular.

Singapore, Singapore
January 2016

Rongxing Lu

Acknowledgments

I am deeply indebted to a number of friends and colleagues in encouraging me to start this work, polish it, and finally publish it. This book would not have been possible without all their efforts in the field, and I'm grateful to my peers and mentors for helping shape my understanding of this area.

The list of people who have lent a hand on this monograph may seem unending, but all names are worthy of mention. I would like to thank the contributors of the presented research works in this monograph: Dr. Sherman Shen, Dr. Xiaodong Lin, Dr. Le Chen, Dr. Haiyong Bao, Dr. Zhenfu Cao, Dr. Xu Li, Dr. Xiaohui Liang, Mr. Cheng Huang, and Mr. Khalid Alharbi.

Special thanks go to my research group members at Nanyang Technological University, Singapore, who help me in proofreading, programming, and file editing. Among them, the names who stand out are Dr. Chang Xu, Mr. Cheng Huang, Mr. Beibei Li, Mr. Hao Hu, and Mr. Shuo Chen.

Special thanks are also due to the staff at Springer Science+Business Media, Susan Lagerstrom-Fife and Jennifer Malat, and Dr. Sherman Shen and Mr. Kuan Zhang at the University of Waterloo, Canada, for their help throughout the publication preparation process.

For those people whose names are not mentioned here, I never forget your kind help. It is my privilege to work and live with so many bright and energetic people in my career.

Finally, this book is a gift to my son Austin who is also the gift to me. My parents and wife have been and will always be a great source of inspiration in my life. This book is also dedicated to them.

Contents

Acronyms

AES	Advanced Encryption Standard
ANOVA	Analysis of Variance
BDH	Bilinear Diffie-Hellman
BGN	Boneh-Goh-Nissim
BiBa	Bins Balls
CC	Control Center
CDH	Computational Diffie-Hellman
CH	Cluster Head
CMs	Cluster Members
DBDH	Decisional Bilinear Diffie-Hellman
DDH	Decisional Diffie-Hellman
DLP	Discrete Logarithm Problem
ε-DP	ε-Differential Privacy
DPAFT	Differentially Private Data Aggregation with Fault Tolerance
DSA	Digital Signature Algorithm
EPPA	Efficient Privacy-Preserving Aggregation
GHG	Greenhouse Gas
GW	Gateway
HAN	Home Area Network
HMAC	Hash-Based Message Authentication Code
HORSE	Hash to Obtain Random Subsets Extension
HPKE	Homomorphic Public Key Encryption
ICT	Information and Communications Technology
IED	Intelligent Electronic Devices
ID	Identity
IND-CPA	Indistinguishable Chosen-Plaintext Attack
JPBC	Java Pairing-Based Cryptography
MAC	Message Authentication Code
MuDA	Multifunctional Data Aggregation
NIST	National Institute of Standards and Technology
OA	Operation Authority

PDF	Probability Density Function
PDAFT	Privacy-Preserving Data Aggregation with Fault Tolerance
PKE	Public Key Encryption
PPMDA	Privacy-Preserving Multidimensional Data Aggregation
PPSDA	Privacy-Preserving Subset Data Aggregation
RA	Residential Area
SGD	Subgroup Decision
SMs	Smart Meters
TA	Trusted Authority

Chapter 1
Introduction

1.1 Background

As modern society moves into the twenty-first century, rapid growth of manufacturing processes, service industries, and household activities require massive demands of electricity supply. It is reported that electricity accounts for around 40 % of the total energy in the United States, and in other countries with similar levels of economic development [1]. In addition, under the drive of the European Union Energy and Climate Package, which has set out targets of reduction in greenhouse gas (GHG) emissions for year 2020 and beyond, governments of many nations have agreed to join in this international agreement to reduce GHG emissions and promote the development of renewable clean resources. What's more, energy losses in the legacy transmission and distribution system have nearly increased twice till 2001 in comparison with that in 1970. Generally speaking, up to 8 % of the electric energy has been lost while transmission and distribution in most current power systems. Energy efficiency has now come to the force as another critical issue.

Most importantly, on August 14, 2003, a cascading system outage of generation and transmission facilities in the North American Eastern Inter-connection caused a blackout of most of New York state as well as parts of Pennsylvania, Ohio, Michigan, and even Ontario, Canada [2]. On November 4, 2006, more than 15 million clients spanning across nearly the whole Europe failed to get access to electricity for about 2 h on this date due to the widespread grid disturbance [3]. The blackouts in 2003 and 2006 explicitly exposed the drawbacks of current electric grid that is lack of sufficient inter-communications and effective real-time diagnosis [2, 3]. It has become apparent that these incidents obviously demonstrate the undoubted obsolescence of the traditional power grid, and it will no longer meet our growing requirements for high security, reliability, availability, and quality of electric supply. Therefore, governors and researchers have begun to explore a future smart grid to satisfy these requirements.

R. Lu, *Privacy-Enhancing Aggregation Techniques for Smart Grid Communications*, Wireless Networks, DOI 10.1007/978-3-319-32899-7_1

1.2 Smart Grid

According to a definition by the Electric Power Research Institute, a smart grid is a power system made up of numerous automated transmission and distribution systems, all operating in a coordinated, efficient and reliable manner. This power system serves millions of customers and has an intelligent communication infrastructure enabling the timely, secure, and updatable information flow needed to provide power to evolving digital economy [4]. The U.S. Energy Independence and Security Act of 2007 directed the National Institute of Standards and Technology (NIST) to coordinate the research and development of a framework to achieve interoperability of smart grid systems and devices. The final full version of the NIST Framework and Roadmap for Smart Grid Interoperability Standards has been released on October 1, 2014 [5]. In this reference model, the smart grid has been divided into seven domains including generation, transmission, distribution, markets, operations, service provider, and customers. The expected benefits and requirements of smart grid are shown below [6]:

- Improving the system reliability and electricity quality;
- Automating system maintenance and operation;
- Enhancing the capacity and efficiency of existing electric power networks;
- Enabling predictive maintenance and self-healing responses to system disturbances;
- Facilitating expanded deployment and integration of renewable energy sources;
- Reducing GHG emissions by enabling electric vehicles and new power sources.

Based on the reference model, we can see that a smart grid system is comprised of several subsystems. It is eventually a network of networks. From a technical view point, the smart grid can be divided into three major systems: smart infrastructure, smart communication and smart management systems.

1.2.1 Smart Infrastructure System

The smart infrastructure system is the physical energy infrastructure integrated with intelligent electronic devices (IEDs) underlying the smart grid. It supports two-way flow of electricity, which implies that the electric energy delivery is not unidirectional anymore. In other words, in the traditional power grid, the electricity is generated by the generation plant, then moved by the transmission and the distribution grid, and finally delivered to the customers. While, in smart grid, electricity can also be transmitted back into the grid by customers, as they may be able to generate electricity using household solar panels and put it back into the microgrid market to sell.

In addition, the smart infrastructure system is also capable of supporting the integration of renewable power resources. These new resources play an important role in providing clean energy and reduce GHG emissions. However, it is a challenging task to combine different kinds of power resources into a centralized power grid. Advanced integration techniques will be utilized in smart grid to address such challenges.

Furthermore, electricity of traditional power system is generated at a centralized power plant, while smart infrastructure system distributes the power generation to several power plants. This can not only improve the robustness and reliability of the power system but also reduce the transmission loss of the electricity.

1.2.2 Smart Communication System

Smart communication system is based on the wide area measurement architecture, wherein smart meters, phasor measurement units are widely deployed across the whole power system to measure customers' detailed electricity usage and the real-time status of the power system. The collected measurement data are then transformed to the phasor data concentrator for aggregation and preprocessing. Eventually, these real-time data are transformed to and analyzed by the system control center for various applications.

The smart communication system constitutes home area network (HAN), local area network (LAN), and wide area network (WAN). The HAN can be connected by mesh points, 3G cellular, WiFi, or WiMax. LAN is usually connected by wired lines such as power line communications, RS-232/RS-485 serial links, or wireless local area networks. The backbone network is undertaken by the WAN, wherein the possible solutions can be dedicated lines or optic fires.

1.2.3 Smart Management System

The smart management system is the most critical subsystem in smart grid that provides advanced real-time management and control services. The key reason that the management system needs to be revolutionized is the lack of real-time analysis and control functionality exposed in the blackouts in 2003 and 2006. The integration of information and communications technologies provides assistance in achieving real-time management. The smart management system will take advantages of the smart infrastructure and smart communication system to pursue various advanced management objectives, such as autonomous system maintenance and operation, self-healing responding to system disturbances, optimal generation and distribution.

In addition, advanced techniques can also be leveraged to identify bad measurement data, fix system failures, detect intrusions, and preserve user privacy. Thus far, such actions are able to achieve high reliability and robustness, energy efficiency improvement, supply and demand balance, emission control, operation cost reduction, and utility maximization.

1.3 Security and Privacy Threats

1.3.1 Security Threats

The future smart grid is expected to enhance the security and reliability of the existing power system. Nevertheless, without strong security and privacy-preserving mechanisms in place, the smart grid will not only inherit the massive existing vulnerabilities of the legacy power grid but also introduce new potential classes of vulnerabilities with the integration of various novel technologies.

Many security threats have been reported for the existing power system so far. In 2010, it was reported that the nuclear power plant in Iran was infected by a network worm named 'Stuxnet' and it spread rapidly to other critical components. This malicious attack has cost huge loss of efforts and money to the nation since then [7]. In addition, in 2008, the Idaho National Laboratory published a report describing several vulnerabilities of the existing power management system [8]. Still, many flaws in the legacy power system and even the smart grid might not have been publicly announced. We categorize the potential attacks into the following two groups: physical attacks and cyber attacks [9].

1.3.1.1 Physical Attacks

As aforementioned, the smart infrastructure system is deployed with a number of IEDs, including smart meters, intelligent appliances, distributed generators, and storage equipment. These devices are located in physically insecure environments, and maintain two-way communications with the electric system, therefore leaving numerous entry points to the power grid. Due to their unprotected locations, it is easy for malicious attackers to exploit the potential vulnerabilities of these devices to either cause local damages or gain access to the more critical parts of the grid by taking advantage of the two-way communications. McLaughlin et al. introduced us on how authentication keys can be extracted from the memory of a smart meter and malicious codes can be inserted into the micro-system to launch attacks via this compromised device [10]. The physical attacks include:

- Physical damage to the IEDs;
- Removal and replacement of the IEDs;
- Manipulation of the communication interface or interruption of the normal communication link between the IED and other devices;
- Cut off of the power transmission lines;
- Electronic disturbance of the electric signal over the transmission lines.

1.3.1.2 Cyber Attacks

Smart grid is integrated with the capabilities of digital communications, controls, self-healing and decision making. There are a number of computer communication networks that can be utilized in smart grid communications. Some of them are wired communication systems while others are wireless systems. The existing computer communication networks are suffering a series of vulnerabilities. Similarly, parallel cyber attacks in computer networks can also be moved to smart grid communication system. These cyber attacks consist of:

- Eavesdropping: The goal of the attacker is to violate the confidentiality of the communication, e.g., by sniffing packets on the LAN or by intercepting wireless transmissions.
- Data tempering: This attacker is able to falsify the exchanged data while transmission, further leading false data analysis.
- Denial of service: The goal of the attacker is to decrease the availability of the system for its intended purpose.
- Spoofing: The attacker tries to disguise as a legitimate partner to participate in the normal communication between two ends to sproof one or two partners.

The smart grid will not only attract normal hackers with various motivations, like unethical customers, publicity seekers, curious or motivated eavesdroppers, but also attract terrorists who aim to disrupt the grid as well, as smart grid is critical infrastructure of nation importance. When many individuals with high motivations and rich resources aim at attacking the system, the risk of finding and exploiting the vulnerabilities and penetrating to the system increases.

1.3.2 Privacy Threats

Apart from the security threats, customers' concern about their privacy is another major obstacle to the public adoption and participation of smart grid. Unlike the legacy power grid in which metering data are read monthly or half monthly, in the smart grid, more detailed and granular energy usage data are collected through smart meters at much shorter time intervals (about every 15 min or less) [11]. Although these precise data are critical to efficient electricity distribution, demand-supply management, optimal power flow analysis, and so on, they might expose a great amount of valuable and intimate information about the customers. It includes the energy usage patterns and the types of household devices and appliances that customers may use, by which the individual activities in a specific household can be clearly deduced [12].

Another challenging issue is the ownership of the valuable collected data. Several entities can benefit from the data including power system control center, electric companies, utility companies, and appliance manufacturers. For example, the Google PowerMeter service [13] can receive real-time usage statistics from installed household smart meters. Customers subscribing to the service receive a customized web page that visualizes local usage. Although Google has yet to announce the final privacy policy for this service, early versions leave the door open to the company using this information for commercial purposes, such as marketing individual or aggregate usage statistics to third parties. Although services such as Google PowerMeter are optional, the customers have less control over the use of power information delivered to utility companies. Therefore, the ownership and accessibility of the confidential data should be identified clearly.

1.4 Security and Privacy Requirements

The reliability, efficiency, and robustness of a smart grid depends on the security of the control and communication systems. In the exploration and development stage of smart grid, communication systems are becoming more and more sophisticated with extensive interconnection, high nonlinearity, and dynamics. Smart grid will require higher degrees of network techniques, meanwhile with higher degree of sophisticated security protocols to deal with the potential vulnerabilities and breaches to support the new features. Here, we first discuss the high level security and privacy requirements in smart grid [14].

- Integrity: The integrity objective refers to preventing undetected modification of information by unauthorized persons or systems. For automation systems, this applies to information such as metering readings, sensor values, or control commands. This objective includes defense against information modification via data injection, data replay, and data delay or drop over the network.
- Confidentiality & Privacy: The confidentiality objective refers to preventing disclosure of information to unauthorized persons or systems. For automation systems, this is relevant both with respect to domain specific information, such as product recipes or plant performance and planning data, and to the secrets specific to the security mechanisms themselves, such as passwords and encryption keys. For customers, privacy issues have to be covered with the derived customer consumption data as they are created in metering devices. Consumption data contains detailed information that can be used to gain insights on a customer's behavior.
- Availability: Availability refers to ensuring that unauthorized persons or systems cannot deny access or use to authorized users. For automation systems, this refers to all the IT elements of the plant, like control systems, safety systems, operator workstations, engineering workstations, manufacturing execution systems, as well as the communication systems between these elements and to the outside world. Violation of availability, also known as denial-of-service (DoS), may not only cause economic damages but may also affect safety issues as operators may lose the ability to monitor and control the process.

- Authentication: Authentication is concerned with determination of the true identity of a system user and mapping of this identity to a system-internal principal (e.g., valid user account) by which this user is known to the system. Most other security objectives, most notably authorization, distinguish between legitimate and illegitimate users based on authentication.
- Authorization: The authorization objective, also known as access control, is concerned with preventing access to the system by persons or systems without permission to do so. In the wider sense, authorization refers to the mechanism that distinguishes between legitimate and illegitimate users for all other security objectives, e.g., confidentiality, integrity, etc. In the narrower sense of access control, it refers to restricting the ability to issue commands to the plant control system. Violation of authorization may cause safety issues.
- Auditability: Auditability is concerned with being able to reconstruct the complete history of the system behavior from historical records of all (relevant) actions executed on it. This security objective is mostly relevant to discover and find reasons for malfunctions in the system after the fact, and to establish the scope of the malfunction or the consequences of a security incident. Note that auditability without authentication may serve diagnostic purposes, but does not provide accountability.
- Nonrepudiability: The nonrepudiability objective refers to being able to provide irrefutable proof to a third party of who initiated a certain action in the system, even if this actor is not cooperating. This security objective is relevant to establish accountability and liability. In the context of automation systems, this is most important with regard to regulatory requirements, e.g., U.S. Food and Drug Administration approval. Violation of this security objective has typically legal/commercial consequences, but no safety implications.

1.5 Challenges

Smart grid is a combination of legacy power system and new information and communication techniques. New standards and regulations need to be embedded into the novel communication network to support the challenges of the future electricity network. The major challenges in building and operating a secure smart grid communication system include access control, privacy, secure operations, efficiency and scalability, and security policies. In this monograph, we focus on customer privacy, and efficiency of data transmission.

1.5.1 User Privacy

With intelligent devices deployed in each household, say smart meter, the detailed usage of the electricity in a specific house is recorded and transmitted to the system control center. While, the detailed usage data can reflect the customer's personal behavior. Therefore, possible leakage or misuse of these data will expose the customer's privacy.

On the one hand, the ownership of these confidential data is one possible issue to take into consideration. Various utilities can make use of such data to make profits for themselves. For example, appliance manufacturers can improve their current products or direct to other novel products to satisfy the customers' meets. As a result, strict and fine-grained access control of the data is an effective way to address such problem.

On the other hand, due to the potential cyber vulnerabilities of the communication system, hackers may temper into the communication link to eavesdrop such data while transmission. This way can also gain access to these confidential data, which might be sold to other profit-related partners in dark market or even launch malicious attacks with these data.

1.5.2 Secure Data Aggregation

In smart grid, a number of meters and sensors are widely deployed across the power system. To achieve real-time monitoring and control, a huge amount of real-time data are collected and reported to the system control center. With the received data, the control center can automatically and timely monitor grid status, balance electricity load, maintain system operation, optimize energy consumption, etc. Specifically, all the intelligent electric appliances in the residential user's home are connected to a smart meter, which periodically records the power consumption of appliances and reports the metering data to a local area substation. The substation then collects, preprocesses (e.g., authenticates, aggregates) and forwards the data to the control center for further analyzing and processing. In addition, phasor measurement units located at each bus will also collect and report the system operating status at a high frequency, say 50 Hz to the control center.

For this problem, a secure aggregation scheme will efficiently solve it. However, as aforementioned, the customer's privacy should also be preserved while transmission. Existing data aggregation schemes also stress the same consideration during aggregating, individual user's data should not be exposed. Most of them use a homomorphic encryption to encrypt users' data so that the semi-trust aggregator can aggregate all users' data without decryption. However, all of these schemes can only be used to compute the summation of users' data as the aggregation, while the CC may need to compute more statistics such as the mostly common used variance. Therefore, how to compute more complex statistics of users' data without disclosing individual user's data is a challenging problem in smart grid communications. Actually, this problem is also mentioned as one of the open research challenges in Internet of Things (IoT).

1.6 Related Activities

To better understand the research challenge of data aggregation in smart grid communications, we first review some existing research efforts made on data aggregation. Data aggregation is a crucial technique in smart grid communications, which is used for collecting the most critical data in an energy efficient manner with minimum data latency. In past decade, many data aggregation schemes have been proposed, including flat network based data aggregation[15, 16] and hierarchal network based data aggregation [17, 18]. However, all these schemes do not consider the security challenges, which results in the possible disclosure of sensitive data, e.g., customer privacy, and the pollution of aggregated data due to malicious devices involved. To address the security issues, Przydatek et al. [19] propose a framework for secure data aggregation, and present an approach called aggregate-commit-prove to enable the data receiver to verify the authenticity of the information provided by the aggregator with efficient random sampling mechanisms and interactive proofs. Although Przydatek et al.'s framework enables secure data aggregation; the effectiveness of the approach still needs to be verified by extensive simulations and real experiments. Meanwhile, data privacy is also not well protected in the framework.

To address the privacy issue in data aggregation, Shi et al. [20] propose the privacy-preserving aggregation of time-series data, which allows a group of nodes to periodically upload encrypted values to a data aggregator, and the latter can calculate the sum of all users' values in every time period, but cannot learn anything else. In [21], Joye and Libert observe that Shi et al.'s protocol just supports small plaintext spaces, and thus present a practical privacy-preserving aggregation protocol to accommodate large plaintext spaces. To support both spatial and temporal aggregation, EKin and Tsudik [22] explore some simple yet relatively efficient privacy techniques for the aggregation of smart meter measurements. Although the above three protocols [20–22] achieve privacy-preserving property, they are not fault-tolerant, that is, once a node is malfunctioning, the whole aggregated results cannot be correctly decrypted. Therefore, fault-tolerant and privacy-preserving aggregation is still challenging. To address this challenge, Chan et al. [23] propose a privacy-preserving stream aggregation with fault tolerance. Though it is resilient to user failure and compromise, the protocol is not quite efficient in terms of transmission cost. In addition to the above privacy-preserving data aggregation works, privacy-preserving data aggregation [24] is also applied to people-centric urban to achieve user privacy, yet the authentication is not well considered. Finally, we should take a note that, homomorphic encryption is an important technique to implement most privacy-preserving aggregation protocols [22], however, since homomorphic encryption is a time-consuming public key technique, it is not efficient for designing privacy-preserving data aggregation, especially for low-cost nodes in smart grid communications.

1.7 Outline of the Book

The organization of the remainder of the monograph is as follows. Chapter 2 presents the homomorphic public key encryption (PKE) techniques, including Paillier PKE and BGN PKE, as the preliminaries. Chapter 3 introduces an efficient privacy-preserving multi-dimensional data aggregation scheme (PPMDA) for secure smart grid communications. A privacy-preserving subset data aggregation (PPSDA) scheme is proposed in Chap. 4. An enhanced multifunctional data aggregation scheme, named MuDA, for privacy-preserving smart grid communications is discussed in Chap. 5. Then, a privacy-preserving data aggregation scheme with fault tolerance, called PDAFT, and a secure enhanced differentially private data aggregation with fault tolerance (DPAFT) are presented in Chaps. 6 and 7, respectively. Finally, Chap. 8 shows a lightweight data aggregation scheme achieving privacy preservation and data integrity with differential privacy and fault tolerance for smart grid communications.

References

1. C. W. Gellings, M. Samotyj, and B. Howe, "The future's smart delivery system [electric power supply]," *IEEE Power Energ. Mag.*, vol. 2, no. 5, pp. 40–48, 2004.
2. G. Andersson, P. Donalek, R. Farmer, N. Hatziargyriou, I. Kamwa, P. Kundur, N. Martins, J. Paserba, P. Pourbeik, and J. Sanchez-Gasca, "Causes of the 2003 major grid blackouts in north america and europe, and recommended means to improve system dynamic performance," *IEEE Trans. Power Syst.*, vol. 20, no. 4, pp. 1922–1928, 2005.
3. G. Maas, M. Bial, and J. Fijalkowski, "Final report-system disturbance on 4 november 2006," *Union for the Coordination of Transmission of Electricity in Europe, Tech. Rep*, 2007.
4. NIST, "Nist framework and roadmap for smart grid interoperability standards," 2010.
5. ——, "Nist releases final version of smart grid framework, update 3.0," 2014.
6. X. Fang, S. Misra, G. L. Xue, and D. J. Yang, "Smart grid - the new and improved power grid: A survey," *Ieee Commun Surv Tut*, vol. 14, no. 4, pp. 944–980, 2012.
7. J. P. Farwell and R. Rohozinski, "Stuxnet and the future of cyber war," *Survival*, vol. 53, no. 1, pp. 23–40, 2011.
8. I. N. Laboratory, "Common cyber security vulnerabilities observed in control system assessments by the inl nstb program," 2008.
9. R. Mahmud, R. Vallakati, A. Mukherjee, P. Ranganathan, and A. Nejadpak, "A survey on smart grid metering infrastructures: Threats and solutions," in *2015 IEEE International Conference on Electro/Information Technology (EIT)*. IEEE, 2015, pp. 386–391.
10. S. McLaughlin, D. Podkuiko, and P. McDaniel, "Energy theft in the advanced metering infrastructure," *Critical Information Infrastructures Security*, vol. 6027, pp. 176–187, 2010.
11. P. Jokar, N. Arianpoo, and V. Leung, "A survey on security issues in smart grids," *Security and Communication Networks*, 2012.
12. C. Laughman, K. Lee, R. Cox, S. Shaw, S. Leeb, L. Norford, and P. Armstrong, "Power signature analysis," *IEEE Power Energ. Mag.*, vol. 1, no. 2, pp. 56–63, 2003.
13. K. Allan, "Power to the people," *Engineering & Technology*, vol. 4, no. 18, pp. 46–49, 2009.
14. D. Dzung, M. Naedele, T. P. Von Hoff, and M. Crevatin, "Security for industrial communication systems," *Proc. IEEE*, vol. 93, no. 6, pp. 1152–1177, 2005.

15. J. Kulik, W. Heinzelman, and H. Balakrishnan, "Negotiation-based protocols for disseminating information in wireless sensor networks," *Wireless networks*, vol. 8, no. 2/3, pp. 169–185, 2002.
16. B. Krishnamachari, D. Estrin, and S. Wicker, "The impact of data aggregation in wireless sensor networks," in *Proceedings of the 22nd International Conference on Distributed Computing Systems Workshops*. IEEE, 2002, pp. 575–578.
17. S. Chatterjea and P. Havinga, "A dynamic data aggregation scheme for wireless sensor networks," 2003.
18. S. Lindsey, C. Raghavendra, and K. M. Sivalingam, "Data gathering algorithms in sensor networks using energy metrics," *IEEE Trans. Parallel Distrib. Syst.*, vol. 13, no. 9, pp. 924–935, 2002.
19. B. Przydatek, D. Song, and A. Perrig, "Sia: Secure information aggregation in sensor networks," in *Proceedings of the 1st International Conference on Embedded Networked Sensor Systems*. ACM, 2003, pp. 255–265.
20. E. Shi, T.-H. H. Chan, E. G. Rieffel, R. Chow, and D. Song, "Privacy-preserving aggregation of time-series data," in *NDSS*, vol. 2, 2011, p. 4.
21. M. Joye and B. Libert, "A scalable scheme for privacy-preserving aggregation of time-series data," in *Financial Cryptography and Data Security*. Springer, 2013, pp. 111–125.
22. Z. Erkin and G. Tsudik, "Private computation of spatial and temporal power consumption with smart meters," in *Applied Cryptography and Network Security*. Springer, 2012, pp. 561–577.
23. T.-H. H. Chan, E. Shi, and D. Song, "Privacy-preserving stream aggregation with fault tolerance," in *Financial Cryptography and Data Security*. Springer, 2012, pp. 200–214.
24. J. Shi, R. Zhang, Y. Z. Liu, and Y. C. Zhang, "Prisense: Privacy-preserving data aggregation in people-centric urban sensing systems," *Ieee Infocom Ser*, 2010.

Chapter 2
Homomorphic Public Key Encryption Techniques

In order to attain privacy-preserving data aggregation in smart grid communications, we need to first understand the homomorphic public key encryption (HPKE) techniques [1–7]. Different from the general public key encryption algorithms [8, 9], HPKE further holds an additional "homomorphic" property, which makes the privacy-preserving data aggregation possible. As shown in Fig. 2.1, when we directly operate over two encrypted data $E(x)$ and $E(y)$ with some operation "$\bullet$", we can gain $E(x \circ y) = E(x) \bullet E(y)$. In most HPKE cases, the operations "$\bullet$" and "$\circ$" are respectively referred to the common multiplication "$\times$" and addition "$+$" operations. Because most of privacy-enhancing aggregation techniques illustrated in this monograph are based on Paillier public key encryption [2] and Boneh-Goh-Nissim (BGN) public key encryption [6], in this chapter, we first take a close look at these two popular homomorphic encryption techniques. Note that, both of them are randomized encryption algorithms [1], i.e., in addition to the plaintext as the input of encryption, a random number is also input for achieving semantic security.

2.1 Paillier Public Key Encryption

Paillier Public Key Encryption (PKE) was first proposed in 1999 [2]. Because of its nice "homomorphic" property, Paillier PKE has received considerable attention and has been widely applied in various privacy-preserving computations. In this section, we will briefly recall the famous homomorphic encryption technique. Before that, we first introduce some basic mathematical backgrounds briefly, which may help the reader digest the Paillier PKE better.

R. Lu, *Privacy-Enhancing Aggregation Techniques for Smart Grid Communications*, Wireless Networks, DOI 10.1007/978-3-319-32899-7_2

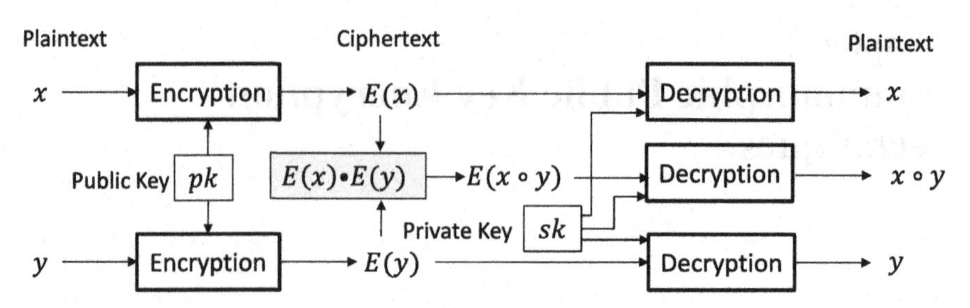

Fig. 2.1 Diagram of homomorphic public key encryption

2.1.1 Mathematical Background

Let $p = 2p' + 1$ and $q = 2q' + 1$ be two safe primes, where p' and q' are also two primes. Compute $n = pq$, we need to prove the following two results:

1. for any $x \in \mathbb{Z}_n$, we have $(1+n)^x = 1 + x \cdot n \bmod n^2$
2. let $\lambda = lcm(p-1, q-1) = 2p'q'$ be the least common multiple of $p-1$ and $q-1$. For any $x \in \mathbb{Z}^*_{n^2}$, we have $x^{n\lambda} = 1 \bmod n^2$.

For the first result, we use the following theorem to prove it.

Theorem 2.1. *For any* $x \in \mathbb{Z}_n$, *we have* $(1+n)^x = 1 + x \cdot n \bmod n^2$.

Proof. When $x = 0$, the result obviously holds. When $0 < x < n$, we have

$$\begin{aligned}(1+n)^x &= \sum_{i=0}^{x} \binom{x}{i} 1^{n-i} \cdot n^i \bmod n^2 \\ &= 1 + n \cdot x + \binom{x}{2} n^2 + \cdots + \binom{x}{x} n^x \bmod n^2 \\ &= 1 + n \cdot x \bmod n^2\end{aligned} \tag{2.1}$$

As a result, the theorem is correct. ■

For the second result, we should use some lemmas and theorem as below.

Definition 2.1 (Euler Totient Function). Let $P = \prod_i p_i^{l_i}$ with p_i pairwise different primes and $l_i > 0$. Then, Euler Totient Function is defined as

$$\phi(P) = P \cdot \prod_i (1 - \frac{1}{p_i}) \tag{2.2}$$

Lemma 2.1. $\phi(p) = p - 1, \phi(q) = q - 1, \phi(n) = (p-1)(q-1), \phi(p^2) = p\phi(p), \phi(q^2) = q\phi(q), \phi(n^2) = n\phi(n)$.

Proof. From the definition of the Euler Totient Function, this lemma can be easily proved. ■

According to the Euler Theorem, for any $x \in \mathbb{Z}_{n^2}^*$, we have $x^{\phi(n^2)} = x^{n\phi(n)} = x^{n \cdot 2\lambda} = 1 \bmod n^2$, but we still cannot determine whether $x^{n\lambda} = 1 \bmod n^2$. In order to obtain $x^{n\lambda} = 1 \bmod n^2$, we need to use the result of the Chinese Remainder Theorem [10].

Theorem 2.2 (Chinese Remainder Theorem). *Suppose that $m_1, m_2, \cdots, m_k$ are pairwise relatively prime positive integers, and let $a_1, a_2, \cdots, a_k$ be integers. Then, the system of congruences, $x \equiv a_i \bmod m_i$ for $1 \le i \le k$, has a unique solution modulo $M = m_1 \times m_2 \times \cdots \times m_k$, which is given by*

$$x \equiv a_1 M_1 y_1 + a_2 M_2 y_2 + \cdots + a_k M_k y_k \bmod M$$

where $M_i = \frac{M}{m_i}$ and $y_i \equiv \frac{1}{M_i} \bmod m_i$ for $1 \le i \le k$.

Let $x \in \mathbb{Z}_{n^2}^*$, from the Euler Theorem, we have

$$\begin{cases} x^{\phi(p^2)} = x^{p(p-1)} = x^{p \cdot 2p'} = 1 \bmod p^2 \Rightarrow x^{pq \cdot 2p'q'} = 1^{qq'} \bmod p^2 \Rightarrow x^{n\lambda} = 1 \bmod p^2 \\ x^{\phi(q^2)} = x^{q(q-1)} = x^{q \cdot 2q'} = 1 \bmod q^2 \Rightarrow x^{pq \cdot 2p'q'} = 1^{pp'} \bmod q^2 \Rightarrow x^{n\lambda} = 1 \bmod q^2 \end{cases} \tag{2.3}$$

Because $\gcd(p^2, q^2) = 1$, we can apply the Extended Euclidean Algorithm to find two integers s, t such that $s \cdot p^2 + t \cdot q^2 = 1$. Let $m_1 = p^2$, $m_2 = q^2$, then $M = m_1 m_2 = n^2$, $M_1 = q^2$, $M_2 = p^2$, $y_1 = t$, and $y_2 = s$. Based on the Chinese Remainder Theorem, where $a_1 = a_2 = 1$, we have

$$x^{n\lambda} = a_1 M_1 y_1 + a_2 M_2 y_2 \bmod M = 1 \cdot q^2 \cdot t + 1 \cdot p^2 \cdot s \bmod n^2 = 1 \bmod n^2 \tag{2.4}$$

Theorem 2.3. *For any $x \in \mathbb{Z}_{n^2}^*$, we have $x^{n\lambda} = 1 \bmod n^2$.*

Definition 2.2 (n-th Residues Modulo n^2). A number $y \in \mathbb{Z}_{n^2}^*$ is said to be an n-th residues modulo n^2 if there exists a number $x \in \mathbb{Z}_{n^2}^*$ such that $y = x^n \bmod n^2$.

Let **NR** be the set of n-th residues modulo n^2. It has been proved that the size of **NR** is exactly $\phi(n)$, i.e., $|\mathbf{NR}| = \phi(n)$. In addition, it has also been proved that "given $y \in \mathbb{Z}_{n^2}^*$, decide whether or not y is n-th residue modulo n^2" is a hard problem, i.e., there does not exist an algorithm that solves the problem in a polynomial time [10].

Let $\mathscr{G} = \{u \in \mathbb{Z}_{n^2}^* | ord(u) = kn, 1 \le k \le \lambda\}$. A special case $g = (1+n)^a \bmod n^2$ with $ord(g) = n$ for some random integer $a \ge 1$ belongs to $\mathscr{G}$, because $g^n = (1 + n)^{an} = 1 \bmod n^2$. Define a function $f(m, r) = g^m r^n \bmod n^2$ over $\mathbb{Z}_n \times \mathbb{Z}_n^* \to \mathbb{Z}_{n^2}^*$, we can show that it is a bijective function. First, because $|\mathbb{Z}_n| = n$, $|\mathbb{Z}_n^*| = \phi(n)$, and $|\mathbb{Z}_{n^2}^*| = \phi(n^2) = n\phi(n)$, we have $|\mathbb{Z}_n \times \mathbb{Z}_n^*| = |\mathbb{Z}_{n^2}^*|$, and thus $\mathbb{Z}_n \times \mathbb{Z}_n^* \to \mathbb{Z}_{n^2}^*$ is injective. On the other hand, if for some $(m_0, r_0), (m_1, r_1) \in \mathbb{Z}_n \times \mathbb{Z}_n^*$ with $f(m_0, r_0) = f(m_1, r_1)$, we have

$$\begin{aligned} & g^{m_0} {r_0}^n = g^{m_1} {r_1}^n \bmod n^2 \Rightarrow g^{m_0 - m_1} {r_0}^n = {r_1}^n \bmod n^2 \\ \Rightarrow & g^{(m_0 - m_1)\lambda} {r_0}^{n\lambda} = {r_1}^{n\lambda} \bmod n^2 \Rightarrow g^{(m_0 - m_1)\lambda} = 1 \bmod n^2 \end{aligned} \tag{2.5}$$

which means $ord(g) | (m_0 - m_1)\lambda$. By choosing $g = (1+n)^a \bmod n^2$ with $ord(g) = n$, we have $n | (m_0 - m_1)\lambda$. Because $\gcd(n, \lambda) = 1$, we have $m_0 - m_1 = 0 \bmod n$. As a result, we have $m_0 = m_1 \bmod n \Rightarrow m_0 = m_1$. Once we have $m_0 = m_1$, we can further obtain ${r_0}^n = {r_1}^n \bmod n^2$ from $g^{m_0 - m_1} {r_0}^n = {r_1}^n \bmod n^2$. It has been easily proved that $f(x) = x^n \bmod n^2$ over $\mathbb{Z}_n^* \rightarrow \mathbf{NR}$ is bijective. Therefore, given ${r_0}^n = {r_1}^n \bmod n^2$, we have $r_0 = r_1$. With this nice bijective function $f(m, r) = g^m r^n \bmod n^2$, the Paillier PKE was proposed [2]. In the following, we describe the details of Paillier PKE.

2.1.2 Description of Paillier PKE

The Paillier PKE can achieve the homomorphic properties, which is mainly comprised of three algorithms: key generation, encryption and decryption.

- *Key Generation:* Given the security parameter κ, two large prime numbers $p = 2p' + 1, q = 2q' + 1$ are first chosen, where $|p| = |q| = \kappa$ and p', q' are also both primes. Then, the RSA modulus $n = pq$ and $\lambda = lcm(p-1, q-1) = 2p'q'$ are computed. Define a function $L(u) = \frac{u-1}{n}$, after choosing a generator $g = (1+n) \in \mathbb{Z}_{n^2}^*$, $\mu = (L(g^\lambda \bmod n^2))^{-1} \bmod n$ is further calculated. Then, the public key is $pk = (n, g)$, and the corresponding private key is $sk = (\lambda, \mu)$.
- *Encryption:* Given a message $m \in \mathbb{Z}_n$, choose a random number $r \in \mathbb{Z}_n^*$, and the ciphertext can be calculated as $c = E(m, r) = g^m \cdot r^n \bmod n^2$.
- *Decryption:* Given the ciphertext $c \in \mathbb{Z}_{n^2}^*$, the corresponding message can be recovered as $m = D(c) = L(c^\lambda \bmod n^2) \cdot \mu \bmod n$.

Correctness.

$$\begin{aligned} m = D(c) = L(c^\lambda \bmod n^2) \cdot \mu \bmod n &= \frac{L(c^\lambda \bmod n^2)}{L(g^\lambda \bmod n^2)} \bmod n \\ = \frac{L((g^m \cdot r^n)^\lambda \bmod n^2)}{L(g^\lambda \bmod n^2)} \bmod n &= \frac{L(g^{m\lambda} \cdot r^{n\lambda} \bmod n^2)}{L(g^\lambda \bmod n^2)} \bmod n \\ = \frac{L(g^{m\lambda} \bmod n^2)}{L(g^\lambda \bmod n^2)} \bmod n &= \frac{L((1+n)^{m\lambda} \bmod n^2)}{L((1+n)^\lambda \bmod n^2)} \bmod n \\ = \frac{L(1 + m\lambda n \bmod n^2)}{L(1 + \lambda n \bmod n^2)} \bmod n &= \frac{m\lambda}{\lambda} \bmod n = m \bmod n = m \end{aligned} \tag{2.6}$$

Security. The Paillier PKE is provably secure against chosen plaintext attack, the detailed security analysis can be referred to [2].

Homomorphic Properties.

1. **Addition.** $E(m_1, r_1) \cdot E(m_2, r_2) = E(m_1 + m_2, r_1 r_2)$

$$E(m_1, r_1) \cdot E(m_2, r_2) = g^{m_1} \cdot {r_1}^n \bmod n^2 \cdot g^{m_2} \cdot {r_2}^n \bmod n^2 = g^{m_1+m_2} \cdot (r_1 r_2)^n \bmod n^2 = E(m_1 + m_2, r_1 r_2)$$

2. **Multiplication.** $E(m_1, r_1)^{m_2} = E(m_1 \cdot m_2, r_1^{m_2})$

$$E(m_1, r_1)^{m_2} = (g^{m_1} \cdot {r_1}^n)^{m_2} \bmod n^2 = g^{m_1 m_2} \cdot {r_1}^{n m_2} \bmod n^2 = E(m_1 m_2, r_1^{m_2})$$

3. **Self-Blinding.** $E(m_1, r_1) \cdot r_2^n \bmod n^2 = E(m_1, r_1 r_2)$

$$E(m_1, r_1) \cdot r_2^n \bmod n^2 = g^{m_1} \cdot r_1 {r_2}^n \bmod n^2 = E(m_1, r_1 r_2)$$

Source Code. The sample java source code of Paillier PKE is available in Appendix 1 in this chapter.

2.2 Boneh-Goh-Nissim (BGN) Public Key Encryption

Paillier PKE is a very popular homomorphic encryption technique, but it does not support the homomorphic multiplication directly over two ciphertexts, which may limit its applications in some scenarios. In order to obtain homomorphic multiplication over two ciphertexts, we recall another famous homomorphic encryption technique - Boneh-Goh-Nissim (BGN) PKE [6] in this section. Before delving into the details, we first recall the bilinear pairing techniques, which serve as the basis of BGN PKE.

2.2.1 Bilinear Pairing Techniques

2.2.1.1 Bilinear Groups of Prime Order

Bilinear pairing is an important cryptographic primitive and has been widely adopted in many positive applications in cryptography [11, 12]. Let $\mathbb{G}$ be a cyclic additive group and $\mathbb{G}_T$ be a cyclic multiplicative group of the same prime order q. We assume that the discrete logarithm problems in both $\mathbb{G}$ and $\mathbb{G}_T$ are hard. A bilinear pairing is a mapping $e : \mathbb{G} \times \mathbb{G} \rightarrow \mathbb{G}_T$ which satisfies the following properties:

1. Bilinearity: for any $P, Q \in \mathbb{G}$ and $a, b \in \mathbb{Z}_q^*$, we have $e(aP, bQ) = e(P, Q)^{ab}$.
2. Non-degeneracy: there exists $P \in \mathbb{G}$ and $Q \in \mathbb{G}$ such that $e(P, Q) \neq 1_{\mathbb{G}_T}$.
3. Computability: there exists an efficient algorithm to compute $e(P, Q)$ for all $P, Q \in \mathbb{G}$.

From Reference [11], we note that such a bilinear pairing may be realized using the modified Weil pairing associated with supersingular elliptic curve.

Definition 2.3 (Bilinear Generator). A bilinear parameter generator $\mathscr{G}en$ is a probability algorithm that takes a security parameter κ as input and outputs a 5-tuple $(q, P, \mathbb{G}, \mathbb{G}_T, e)$, where q is a κ-bit prime number, $(\mathbb{G}, +)$ and $(\mathbb{G}_T, \times)$ are two groups with the same order q, $P \in \mathbb{G}$ is a generator, and $e : \mathbb{G} \times \mathbb{G} \rightarrow \mathbb{G}_T$ is an admissible bilinear map.

In the following, we define the quantitative notion of the complexity assumptions, including Computational Diffie-Hellman (CDH) Problem, Decisional Diffie-Hellman (DDH) Problem, and Bilinear Diffie-Hellman (BDH) Problem.

Definition 2.4 (Computational Diffie-Hellman (CDH) Problem). The Computational Diffie-Hellman (CDH) problem in $\mathbb{G}$ is defined as follows: Given P, aP, $bP \in \mathbb{G}$ for unknown $a, b \in \mathbb{Z}_q^*$, compute $abP \in \mathbb{G}$.

Definition 2.5 (CDH Assumption). Let $\mathscr{A}$ be an adversary that takes an input of $(P, aP, bP) \in \mathbb{G}$ for unknown $a, b \in \mathbb{Z}_q^*$, and returns abP. We consider the following random experiment.

$$
\begin{array}{l}
\texttt{Experiment}\ \mathbf{Exp}_{\mathscr{A}}^{\mathrm{CDH}} \\
\quad a, b \stackrel{R}{\leftarrow} \mathbb{Z}_q, \alpha \leftarrow \mathscr{A}(P, aP, bP) \\
\quad \texttt{if}\ \alpha = abP,\ \texttt{then}\ \beta \leftarrow 1,\ \texttt{else}\ \beta \leftarrow 0 \\
\quad \texttt{return}\ \beta
\end{array}
$$

We define the corresponding success probability of $\mathscr{A}$ in solving the CDH problem via

$$\mathbf{Succ}_{\mathscr{A}}^{\mathrm{CDH}} = \Pr\left[\mathbf{Exp}_{\mathscr{A}}^{\mathrm{CDH}} = 1\right]$$

Let $\tau \in \mathbb{N}$ and $\epsilon \in [0, 1]$. We say that the CDH is (τ, ϵ)-secure if no polynomial algorithm $\mathscr{A}$ running in time τ has success $\mathbf{Succ}_{\mathscr{A}}^{\mathrm{CDH}} \geq \epsilon$.

Definition 2.6 (Decisional Diffie-Hellman (DDH) Problem). For $a, b, c \in \mathbb{Z}_q^*$, given P, aP, bP, $cP \in \mathbb{G}$, decide whether $c = ab \in \mathbb{Z}_q$. The DDH problem is easy in $\mathbb{G}$, since we can compute $e(aP, bP) = e(P, P)^{ab}$ and decide whether $e(P, P)^{ab} = e(P, P)^c$ [11].

Definition 2.7 (Bilinear Diffie-Hellman (BDH) Problem). The Bilinear Diffie-Hellman (BDH) problem in $\mathbb{G}$ is as follows: Given $P, aP, bP, cP \in \mathbb{G}$ for unknown $a, b, c \in \mathbb{Z}_q^*$, compute $e(P, P)^{abc} \in \mathbb{G}_T$.

Definition 2.8 (BDH Assumption). Let $\mathscr{A}$ be an adversary that takes an input of $(P, aP, bP, cP) \in \mathbb{G}$ for unknown $a, b, c \in \mathbb{Z}_q^*$, and returns $e(P, P)^{abc}$. We consider the following random experiment.

$$
\begin{aligned}
&\texttt{Experiment}\ \mathbf{Exp}_{\mathscr{A}}^{\mathrm{CDH}} \\
&\quad a, b, c \xleftarrow{R} \mathbb{Z}_q, \alpha \leftarrow \mathscr{A}\,(P, aP, bP, cP) \\
&\quad \texttt{if}\ \alpha = e(P, P)^{abc}\ \texttt{then}\ \beta \leftarrow 1\ \texttt{else}\ \beta \leftarrow 0 \\
&\quad \texttt{return}\ \beta
\end{aligned}
$$

We define the corresponding success probability of $\mathscr{A}$ in solving the BDH problem via

$$\mathbf{Succ}_{\mathscr{A}}^{\mathrm{BDH}} = \Pr\left[\mathbf{Exp}_{\mathscr{A}}^{\mathrm{BDH}} = 1\right]$$

Let $\tau \in \mathbb{N}$ and $\epsilon \in [0, 1]$. We say that the BDH is (τ, ϵ)-secure if no polynomial algorithm $\mathscr{A}$ running in time τ has success $\mathbf{Succ}_{\mathscr{A}}^{\mathrm{BDH}} \geq \epsilon$.

Definition 2.9 (Decisional Diffie-Hellman (DBDH) Problem). The Decisional Bilinear Diffie-Hellman (DBDH) problem in $\mathbb{G}$ is as follows: Given an element P of $\mathbb{G}$, a tuple (aP, bP, cP, T) for unknown $a, b, c \in \mathbb{Z}_q^*$ and $T \in \mathbb{G}_T$, decide whether $T = e(P, P)^{abc}$ or a random element R drawn from $\mathbb{G}_T$.

Definition 2.10 (DBDH Assumption). Let $\mathscr{A}$ be an adversary that takes an input of (aP, bP, cP, T) for unknown $a, b, c \in \mathbb{Z}_q^*$ and $T \in \mathbb{G}_T$, and returns a bit $\beta' \in \{0, 1\}$. We consider the following random experiments.

$$
\begin{aligned}
&\texttt{Experiment}\ \mathbf{Exp}_{\mathscr{A}}^{\mathrm{DBDH}} \\
&\quad a, b, c \xleftarrow{R} \mathbb{Z}_q; R \xleftarrow{R} \mathbb{G}_T \\
&\quad \tilde{\beta} \leftarrow \{0, 1\} \\
&\quad \text{if } \tilde{\beta} = 0, \text{ then } T = e(P, P)^{abc}; \text{ else if } \tilde{\beta} = 1 \text{ then } T = R \\
&\quad \tilde{\beta}' \leftarrow \mathscr{A}(aP, bP, cP, T) \\
&\quad \texttt{return}\ 1 \text{ if } \tilde{\beta}' = \tilde{\beta}, 0 \text{ otherwise}
\end{aligned}
$$

We then define the advantage of $\mathscr{A}$ via

$$
\begin{aligned}
\mathbf{Adv}_{\mathscr{A}}^{\mathrm{DBDH}} = \Big|&\Pr\left[\mathbf{Exp}_{\mathscr{A}}^{\mathrm{DBDH}} = 1 | \tilde{\beta} = 0\right] \\
&- \Pr\left[\mathbf{Exp}_{\mathscr{A}}^{\mathrm{DBDH}} = 1 | \tilde{\beta} = 1\right]\Big| \geq \epsilon
\end{aligned}
$$

Let $\tau \in \mathbb{N}$ and $\epsilon \in [0, 1]$. We say that the DBDH is (τ, ϵ)-secure if no adversary $\mathscr{A}$ running in time τ has an advantage $\mathbf{Adv}_{\mathscr{A}}^{\mathrm{DBDH}} \geq \epsilon$.

2.2.1.2 Bilinear Groups of Composite Order

Let p, q be two distinct large primes, and $n = pq$. Groups $(\mathbb{G}, \mathbb{G}_T)$ of composite order n are called *bilinear map groups of composite order* if there is a mapping $e : \mathbb{G} \times \mathbb{G} \to \mathbb{G}_T$ with the following properties [6, 13]:

1. Bilinearity: $e(g^a, h^b) = e(g, h)^{ab}$ for any $(g, h) \in \mathbb{G}^2$ and $a, b \in \mathbb{Z}_n$.
2. Non-degeneracy: there exists $g \in \mathbb{G}$ such that $e(g, g)$ has order n in $\mathbb{G}_T$. In other words, $e(g, g)$ is a generator of $\mathbb{G}_T$, whereas g generates $\mathbb{G}$.
3. Computability: there exists an efficient algorithm to compute $e(g, h) \in \mathbb{G}_T$ for all $(g, h) \in \mathbb{G}$.

Note that 1) we use the multiplicative group to represent the group $\mathbb{G}$, which, however, can be instantiated by the elliptic curve addition group, i.e., the modified Weil pairing or Tate pairing [6, 13]; 2) the vast majority of cryptosystems based on pairings assume for simplicity that bilinear groups have prime order q. In composite order case, it is important that the pairing is defined over a group $\mathbb{G}$ containing $|\mathbb{G}| = n$ elements, where $n = pq$ has a (ostensibly hidden) factorization in two large primes, $p \neq q$; 3) those complexity assumptions above in bilinear group of prime order also hold in bilinear group of composite order.

Definition 2.11 (Composite Bilinear Generator). A composite bilinear parameter generator $\mathcal{CG}en$ is a probabilistic algorithm that takes a security parameter k as input, and outputs a 5-tuple $(n, g, \mathbb{G}, \mathbb{G}_T, e)$, where $n = pq$ and p, q are two k-bit prime numbers, $\mathbb{G}, \mathbb{G}_T$ are two groups with order n, $g \in \mathbb{G}$ is a generator, and $e : \mathbb{G} \times \mathbb{G} \to \mathbb{G}_T$ is a non-degenerated and efficiently computable bilinear map.

Let $\mathbf{g}$ be a generator of $\mathbb{G}$, then $g = \mathbf{g}^q \in \mathbb{G}$ can generate the subgroup $\mathbb{G}_p = \{g^0, g^1, \cdots, g^{p-1}\}$ of order p, and $g' = \mathbf{g}^p \in \mathbb{G}$ can generate the subgroup $\mathbb{G}_q = \{g'^0, g'^1, \cdots, g'^{q-1}\}$ of order q in $\mathbb{G}$. In the following, we define the quantitative notion of the complexity of the SubGroup Decision (SGD) Problem [6].

Definition 2.12 (SubGroup Decision (SGD) Problem). The SubGroup Decision (SGD) problem in $\mathbb{G}$ is as follows: Given a tuple $(e, \mathbb{G}, \mathbb{G}_T, n, h)$, where the element h is randomly drawn from either $\mathbb{G}$ or subgroup $\mathbb{G}_q$, decide whether or not $h \in \mathbb{G}_q$.

Definition 2.13 (SGD Assumption). Let $\mathcal{A}$ be an adversary that takes an input of h drawn from either $\mathbb{G}$ or subgroup $\mathbb{G}_q$, and returns a bit $b' \in \{0, 1\}$. We consider the following random experiments.

`Experiment` $\mathbf{Exp}_{\mathcal{A}}^{\mathrm{SGD}}$
$\tilde{b} \leftarrow \{0, 1\}$
if $\tilde{b} = 0$, then $h \stackrel{R}{\leftarrow} \mathbb{G}_q$; else if $\tilde{b} = 1$ then $h \stackrel{R}{\leftarrow} \mathbb{G}$
$\tilde{b}' \leftarrow \mathcal{A}(e, \mathbb{G}, \mathbb{G}_T, n, h)$
`return` 1 if $\tilde{b}' = \tilde{b}$, 0 otherwise

We then define the advantage of $\mathscr{A}$ via

$$\mathbf{Adv}_{\mathscr{A}}^{\mathrm{SGD}} = \left|\Pr\left[\mathbf{Exp}_{\mathscr{A}}^{\mathrm{SGD}} = 1|\tilde{b} = 0\right] - \Pr\left[\mathbf{Exp}_{\mathscr{A}}^{\mathrm{SGD}} = 1|\tilde{b} = 1\right]\right| \geq \epsilon$$

Let $\tau \in \mathbb{N}$ and $\epsilon \in [0, 1]$. We say that the SGD is (τ, ϵ)-secure if no adversary $\mathscr{A}$ running in time τ has an advantage $\mathbf{Adv}_{\mathscr{A}}^{\mathrm{SGD}} \geq \epsilon$.

2.2.2 Description of BGN PKE

BGN PKE [6] can achieve one more homomorphic property in comparison to the Paillier PKE, which mainly consists of three algorithms: key generation, encryption, and decryption.

- *Key Generation:* Given the security parameter k, composite bilinear parameters $(n, g, \mathbb{G}, \mathbb{G}_T, e)$ are generated by $\mathscr{C}\mathscr{G}en(k)$, where $n = pq$ and p, q are two k-bit prime numbers, and $g \in \mathbb{G}$ is a generator of order n. Set $h = g^q$, then h is a random generator of the subgroup of $\mathbb{G}$ of order p. The public key is $pk = (n, \mathbb{G}, \mathbb{G}_T, e, g, h)$, and the corresponding private key is $sk = p$.
- *Encryption:* We assume the message space consists of integers in the set $\{0, 1, \cdots, T\}$ with $T < q$. To encrypt a message m, we choose a random number $r \in \mathbb{Z}_n$, and compute the ciphertext $c = E(m, r) = g^m h^r \in \mathbb{G}$.
- *Decryption:* Given the ciphertext $c = E(m, r) = g^m h^r \in \mathbb{G}$, the corresponding message can be recovered by the private key p. Observe that $c^p = (g^m h^r)^p = (g^p)^m$. Let $\hat{g} = g^p$. To recover m, it suffices to compute the discrete log of c^p base $\hat{g}$. Since $0 \leq m \leq T$, the expected time is around $O(\sqrt{T})$ when using the Pollard's lambda method [14](p. 128).

Security. BGN PKE is provably secure against chosen plaintext attack based on the subgroup decision assumption, the detailed security analysis can be referred to [6].

Homomorphic Properties.

1. **Addition.** $E(m_1, r_1) \cdot E(m_2, r_2) = E(m_1 + m_2, r_1 + r_2)$

$$E(m_1, r_1) \cdot E(m_2, r_2) = g^{m_1} h^{r_1} \cdot g^{m_2} h^{r_2} = g^{m_1+m_2} \cdot h^{r_1+r_2} = E(m_1 + m_2, r_1 + r_2)$$

2. **Multiplication.** $E(m_1, r_1)^{m_2} = E(m_1 \cdot m_2, r_1 \cdot m_2)$

$$E(m_1, r_1)^{m_2} = (g^{m_1} \cdot h^{r_1})^{m_2} = g^{m_1 m_2} \cdot h^{r_1 m_2} = E(m_1 m_2, r_1 m_2)$$

3. **Self-Blinding.** $E(m_1, r_1) \cdot h^{r_2} = E(m_1, r_1 + r_2)$

$$E(m_1, r_1) \cdot h^{r_2} = g^{m_1} \cdot h^{r_1+r_2} = E(m_1, r_1 + r_2)$$

4. **Multiplication-II.** $e(E(m_1, r_1), E(m_2, r_2)) = E'(m_1 \cdot m_2, m_1 r_2 + r_1 m_2 + q r_1 r_2)$

$$C = e(E(m_1, r_1), E(m_2, r_2)) = e(g^{m_1} h^{r_1}, g^{m_2} h^{r_2}) = e(g, g)^{m_1 m_2} \cdot e(g, h)^{m_1 r_2 + r_1 m_2 + q r_1 r_2}$$
$$= E'(m_1 \cdot m_2, m_1 r_2 + r_1 m_2 + q r_1 r_2)$$

Observe that $C^p = (e(g, g)^{m_1 m_2} \cdot e(g, h)^{m_1 r_2 + r_1 m_2 + q r_1 r_2})^p = (e(g, g)^p)^{m_1 m_2}$. Let $\overline{g} = e(g, g)^p$. To recover $m_1 m_2$, it suffices to compute the discrete log of C^p base $\overline{g}$ by using the Pollard's lambda method [14](p. 128). Note that the homomorphic multiplication can be taken only *once* upon two ciphertexts in $\mathbb{G}$, and then the result will be in $\mathbb{G}_T$, but it still supports additive homomorphism. Also note that, if we do not expect the **Multiplication-II** homomorphic property in some scenarios, we do not need to use bilinear groups of composite order, and can simply build the BGN PKE over the general group $(\mathbb{G}, \times)$ with composite order $n = pq$.

Source Code. The sample java source code of BGN PKE is available in Appendix 2 in this chapter.

2.3 Summary

In this chapter, we have discussed two popular homomorphic encryption techniques Paillier PKE [2] and BGN PKE [6], which will be used in the design of most privacy-preserving aggregation schemes in this monograph. Note that, the fully homomorphic encryption techniques [15–21] can also be applied in privacy-preserving data aggregation in smart grid communications [22]. However, the efficiency needs to be extensively exploited in practical scenarios. Therefore, in this monograph, the fully homomorphic encryption techniques are not our focuses, interested readers can refer to [15–21] for more details.

Appendix 1: A Sample Java Source Code of Paillier PKE

```
import java.math.BigInteger;
import java.security.SecureRandom;

/**
 * @ClassName: Paillier
 * @Description: This is a sample java source code of Paillier
```

(continued)

```
 * PKE.
 */
public class Paillier {

    /**
     * @ClassName: PublicKey
     * @Description: This is a class for storing the public
     *               key (n, g) of Paillier PKE.
     */
    public class PublicKey {
        private BigInteger n, g;

        public PublicKey(BigInteger n, BigInteger g) {
            this.n = n;
            this.g = g;
        }

        public BigInteger getN() {
            return n;
        }

        public BigInteger getG() {
            return g;
        }
    }

    /**
     * @ClassName: PrivateKey
     * @Description: This is a class for storing the private
     *               key (lambda, mu) of Paillier PKE.
     */
    public class PrivateKey {
        private BigInteger lambda, mu;

        public PrivateKey(BigInteger lambda, BigInteger mu) {
            this.lambda = lambda;
            this.mu = mu;
        }

        public BigInteger getLambda() {
            return lambda;
        }

        public BigInteger getMu() {
            return mu;
        }
    }

    private final int CERTAINTY = 64;
```

(continued)

```
private PublicKey pubkey; // The public key of Paillier
                             PKE, (n, g)
private PrivateKey prikey; // The private key of Paillier
                              PKE, (lambda, mu)
/**
 * @Title: getPubkey
 * @Description: This function returns the generated
 *               public key.
 * @return PublicKey The public key used to encrypt
 * the data.
 */
public PublicKey getPubkey() {
    return pubkey;
}

/**
 * @Title: getPrikey
 * @Description: This function returns the generated
 *               private key.
 * @return PrivateKey The private key used to decrypt
 * the data.
 */
public PrivateKey getPrikey() {
    return prikey;
}

/**
 * @Title: keyGeneration
 * @Description: This function is to help generate the
 *               public key and
 *               private key for encryption and decryption.
 * @param k
 *            k is the security parameter, which decides
 *            the length of two large primes (p and q).
 * @return void
 */
public void keyGeneration(int k) {

    BigInteger p_prime, q_prime, p, q;

    do {
        p_prime = new BigInteger(k, CERTAINTY,
                  new SecureRandom());
        p = (p_prime.multiply(BigInteger.valueOf(2)))
            .add(BigInteger.ONE);
    } while (!p.isProbablePrime(CERTAINTY));

    do {
        do {
            q_prime = new BigInteger(k, CERTAINTY,
```

(continued)

```
                        new SecureRandom());
            } while (p_prime.compareTo(q_prime) == 0);
            q = (q_prime.multiply(BigInteger.valueOf(2)))
                .add(BigInteger.ONE);
        } while (!q.isProbablePrime(CERTAINTY));

        // The following steps are to generate the keys
        // n=p*q
        BigInteger n = p.multiply(q);
        // nsquare=n^2
        BigInteger nsquare = n.pow(2);
        // a generator g=(1+n) in Z*_(n^2)
        BigInteger g = BigInteger.ONE.add(n);
        // lambda = lcm(p-1, q-1) = p_prime*q_prime
        BigInteger lambda = BigInteger.valueOf(2)
                            .multiply(p_prime)
                .multiply(q_prime);
        // mu = (L(g^lambda mod n^2))^{-1} mod n
        BigInteger mu = Lfunction(g.modPow(lambda, nsquare), n)
                        .modInverse(n);

        pubkey = new PublicKey(n, g);
        prikey = new PrivateKey(lambda, mu);
    }

    /**
     * @Title: encrypt
     * @Description: This function is to encrypt the message
     *               with Paillier's public key.
     * @param m
     *           The message.
     * @param pubkey
     *           The public key of Paillier PKE.
     * @return BigInteger The ciphertext.
     * @throws Exception
     *            If the message is not in Z*_n, there is
     *            an exception.
     */
    public static BigInteger encrypt(BigInteger m,
            PublicKey pubkey) throws Exception {
        BigInteger n = pubkey.getN();
        BigInteger nsquare = n.pow(2);
        BigInteger g = pubkey.getG();
        if (!belongToZStarN(m, n)) {
            throw new Exception(
                    "Paillier.encrypt(BigInteger m, PublicKey
                              pubkey): plaintext m is not
                                       in Z*_n");
        }
        BigInteger r = randomZStarN(n);
```

(continued)

```
    return (g.modPow(m, nsquare).multiply(r.modPow(n,
              nsquare))).mod(nsquare);
}

/**
 * @Title: decrypt
 * @Description: This function is to decrypt the ciphertext
 *              with the public key and the private key.
 * @param c
 *           The ciphertext.
 * @param pubkey
 *           The public key of Paillier PKE.
 * @param prikey
 *           The private key of Paillier PKE.
 * @return BigInteger The plaintext.
 * @throws Exception
 *            If the cipher is not in Z*_(n^2), there is
 *            an exception.
 */
public static BigInteger decrypt(BigInteger c, PublicKey
         pubkey, PrivateKey prikey) throws Exception {
    BigInteger n = pubkey.getN();
    BigInteger nsquare = n.pow(2);
    BigInteger lambda = prikey.getLambda();
    BigInteger mu = prikey.getMu();
    if (!belongToZStarNSquare(c, nsquare)) {
        throw new Exception(
            "Paillier.decrypt(BigInteger c, PrivateKey
             prikey): ciphertext c is not in Z*_(n^2)");
    }

    return Lfunction(c.modPow(lambda, nsquare), n).
                      multiply(mu).mod(n);
}

/**
 * @Title: add
 * @Description: The function supports the homomorphic
 * addition with two ciphertext.
 * @param c1
 *           The ciphertext.
 * @param c2
 *           The ciphertext.
 * @param pubkey
 *           The public key of Paillier PKE.
 * @return BigInteger The return value is c1*c2 mod n^2.
 */
public static BigInteger add(BigInteger c1, BigInteger c2,
    PublicKey pubkey) {BigInteger nsquare = pubkey.getN()
    .pow(2);
```

(continued)

```
        return c1.multiply(c2).mod(nsquare);
    }

    /**
     * @Title: mul
     * @Description: The function supports the homomorphic
     *  multiplication with one ciphertext and one plaintext.
     * @param c
     *            The ciphertext.
     * @param m
     *            The plaintext.
     * @param pubkey
     *            The public key of Paillier PKE.
     * @return BigInteger The return value is c^m mod n^2.
     */
    public static BigInteger mul(BigInteger c, BigInteger m,
    PublicKey pubkey) {BigInteger nsquare =
                        pubkey.getN().pow(2);
        return c.modPow(m, nsquare);
    }

    /**
     * @Title: selfBlind
     * @Description: The function supports the homomorphic
     * self-blinding with one ciphertext and one random number.
     * @param c
     *            The ciphertext.
     * @param r
     *            A random number in Z*_n.
     * @param pubkey
     *            The public key of Paillier PKE.
     * @return BigInteger The return value is c*r^n mod n^2.
     */
    public static BigInteger selfBlind(BigInteger c,
            BigInteger r, PublicKey pubkey) {
        BigInteger n = pubkey.getN();
        BigInteger nsquare = n.pow(2);
        return c.multiply(r.modPow(n, nsquare)).mod(nsquare);
    }

    /**
     * @Title: Lfunction
     * @Description: This function is the L function which is
     * defined by Paillier PKE, L(mu)=(mu-1)/n.
     * @param mu
     *            The input parameter.
     * @param n
     *            n=p*q.
     * @return BigInteger The return value is (mu-1)/n.
     */
```

(continued)

```
private static BigInteger Lfunction(BigInteger mu,
 BigInteger n) {return
                mu.subtract(BigInteger.ONE).divide(n);
}

/**
 * @Title: randomZStarN
 * @Description: This function returns a ramdom number in
 *                Z*_n.
 * @param n
 *             n=p*q.
 * @return BigInteger A random number in Z*_n.
 */
public static BigInteger randomZStarN(BigInteger n) {
    BigInteger r;
    do {
        r = new BigInteger(n.bitLength(), new
            SecureRandom());
    } while (r.compareTo(n) >= 0 || r.gcd(n).intValue()
     != 1);
     return r;
}

/**
 *
 * @Title: belongToZStarN
 * @Description: This function is to test whether the
 *                plaintext is in Z*_n.
 * @param m
 *             The plaintext.
 * @param n
 *             n=p*q.
 * @return boolean If it is true, the plaintext is Z*_n,
 *         otherwise, not.
 */
private static boolean belongToZStarN(BigInteger m,
    BigInteger n) {
    if (m.compareTo(BigInteger.ZERO) < 0 ||
    m.compareTo(n) >= 0
            || m.gcd(n).intValue() != 1) {
        return false;
    }
    return true;
}

/**
 *
 * @Title: belongToZStarNSquare
 * @Description: This function is to test whether the
 *                ciphertext is in
```

(continued)

```
     *                Z*_(n^2).
     * @param c
     *            The ciphertext.
     * @param nsquare
     *            nsquare=n^2.
     * @return boolean If it is true, the ciphertext is
     *         Z*_(n^2), otherwise, not.
     */
    private static boolean belongToZStarNSquare(BigInteger c,
        BigInteger nsquare){
        if (c.compareTo(BigInteger.ZERO) < 0 ||
        c.compareTo(nsquare) >= 0
                || c.gcd(nsquare).intValue() != 1) {
            return false;
        }
        return true;
    }

    public static void main(String[] args) {
        Paillier paillier = new Paillier();

        // KeyGeneration
        paillier.keyGeneration(512);
        Paillier.PublicKey pubkey = paillier.getPubkey();
        Paillier.PrivateKey prikey = paillier.getPrikey();

        // Encryption and Decryption
        BigInteger m = new BigInteger(new
                        String("Hello").getBytes());
        BigInteger c = null;
        BigInteger decrypted_m = null;
        try {
            c = Paillier.encrypt(m, pubkey);
            decrypted_m = Paillier.decrypt(c, pubkey, prikey);
        } catch (Exception e) {
            // TODO Auto-generated catch block
            e.printStackTrace();
        }

        if (decrypted_m.compareTo(m) == 0) {
            System.out.println("Encryption and Decryption
            test successfully.");
        }

        // Homomorphic Properties

        // Addition
        BigInteger m1 = new BigInteger("12345");
        BigInteger m2 = new BigInteger("56789");
        BigInteger m1plusm2 = m1.add(m2);
```

(continued)

```
        try {
            BigInteger c1 = Paillier.encrypt(m1, pubkey);
            BigInteger c2 = Paillier.encrypt(m2, pubkey);
            BigInteger c1mulc2 = Paillier.add(c1, c2, pubkey);
            BigInteger decrypted_c1mulc2 =
                        Paillier.decrypt(c1mulc2, pubkey, prikey);
            if (decrypted_c1mulc2.compareTo(m1plusm2) == 0) {
                System.out.println("Homomorphic addition
                tests successfully.");
            }
        } catch (Exception e) {
            e.printStackTrace();
        }

        // Multiplication
        m1 = new BigInteger("12345");
        m2 = new BigInteger("56789");
        BigInteger m1mulm2 = m1.multiply(m2);
        try {
            BigInteger c1 = Paillier.encrypt(m1, pubkey);
            BigInteger c1expm2 = Paillier.mul(c1, m2, pubkey);
            BigInteger decrypted_c1expm2 =
                       Paillier.decrypt(c1expm2, pubkey, prikey);
            if (decrypted_c1expm2.compareTo(m1mulm2) == 0) {
                System.out
                        .println("Homomorphic multiplication
                         tests successfully.");
            }
        } catch (Exception e) {
            e.printStackTrace();
        }

        // Self-Blinding
        m1 = new BigInteger("12345");
        BigInteger r2 = Paillier.randomZStarN(pubkey.getN());
        try {
            BigInteger c1 = Paillier.encrypt(m1, pubkey);
            BigInteger c1mulrn = Paillier.selfBlind(c1, r2,
                         pubkey);
            BigInteger decrypted_c1mulrn =
                       Paillier.decrypt(c1mulrn, pubkey, prikey);
            if (decrypted_c1mulrn.compareTo(m1) == 0) {
                System.out
                        .println("Homomorphic self-blinding
                         tests successfully.");
            }
        } catch (Exception e) {
            e.printStackTrace();
        }
    }
}
```

Appendix 2: A Sample Java Source Code of BGN PKE

```
import java.math.BigInteger;

/*
 * This source code uses the JPBC (Java Pairing-Based
 *      Cryptography) library,
 * which can be downloaded from
 *     http://gas.dia.unisa.it/projects/jpbc/
 */

import it.unisa.dia.gas.jpbc.*;
import it.unisa.dia.gas.plaf.jpbc.pairing.PairingFactory;
import it.unisa.dia.gas.plaf.jpbc.pairing.a1
       .TypeA1CurveGenerator;

/**
 * @ClassName: BGN
 * @Description: This is a sample java source code of BGN PKE.
 */
public class BGN {
    /**
     * @ClassName: PublicKey
     * @Description: This is a class for storing the
     * public key (n,G,GT,e,g,h) of BGN PKE.
     */
    public class PublicKey {
        private BigInteger n;
        private Field<Element> Field_G, Field_GT;
        private Pairing pairing;
        private Element g, h;

        public PublicKey(BigInteger n, Field<Element> G,
                Field<Element> GT, Pairing pairing, Element g,
                Element h) {
            this.n = n;
            this.Field_G = G;
            this.Field_GT = GT;
            this.pairing = pairing;
            this.g = g;
            this.h = h;
        }

        public Element getG() {
            return g;
        }
```

(continued)

```
    public Element getH() {
        return h;
    }

    public BigInteger getN() {
        return n;
    }

    public Pairing getPairing() {
        return pairing;
    }

    public Field<Element> getField_G() {
        return Field_G;
    }

    public Field<Element> getField_GT() {
        return Field_GT;
    }
}

/**
 * @ClassName: PrivateKey
 * @Description: This is a class for storing the
 *               private key (p) of BGN PKE.
 */
public class PrivateKey {
    private BigInteger p;

    public PrivateKey(BigInteger p) {
        this.p = p;
    }

    public BigInteger getP() {
        return p;
    }
}

private static final int T = 100; // The max range of
   message m
private PublicKey pubkey;
private PrivateKey prikey;

/**
 * @Title: keyGeneration
 * @Description: This function is responsible for
 * generating the public keys and the private keys.
 * @param k
 *            the security parameter, which decides the
 *            length of two large prime (p and q).
```

(continued)

```
     * @return void
     */
    public void keyGeneration(int k) {
        TypeA1CurveGenerator pg = new
          TypeA1CurveGenerator(2, k);
        PairingParameters pp = pg.generate();
        Pairing pairing = PairingFactory.getPairing(pp);
        BigInteger n = pp.getBigInteger("n");
        BigInteger q = pp.getBigInteger("n0");
        BigInteger p = pp.getBigInteger("n1");
        Field<Element> Field_G = pairing.getG1();
        Field<Element> Field_GT = pairing.getGT();
        Element g = Field_G.newRandomElement().getImmutable();
        Element h = g.pow(q).getImmutable();

        pubkey = new PublicKey(n, Field_G, Field_GT,
                 pairing, g, h);
        prikey = new PrivateKey(p);
    }

    /**
     * @Title: getPubkey
     * @Description: This function returns the public key of
     *               BGN PKE.
     * @return PublicKey The public key used to encrypt
     *               the data.
     */
    public PublicKey getPubkey() {
        return pubkey;
    }

    /**
     * @Title: getPrikey
     * @Description: This function returns the private key of
     *               BGN PKE.
     * @return PrivateKey The private key used to decrypt
     *               the data.
     */
    public PrivateKey getPrikey() {
        return prikey;
    }

    /**
     * @Title: encrypt
     * @Description: This function is to encrypt the message
     *               m, m in [0,1,2,...,T],
     *               T=100 with public key.
     * @param m
     *            The message
     * @param pubkey
```

(continued)

```
 *              The public key of BGN PKE.
 * @return Element The ciphertext.
 * @throws Exception
 *               If the plaintext is not in [0,1,2,...,n],
 *               there is an exception.
 */
public static Element encrypt(int m, PublicKey pubkey)
        throws Exception {
    if (m > T) {
        throw new Exception(
                "BGN.encrypt(int m, PublicKey pubkey): "
                + "plaintext m is not in [0,1,2,...,"
                        + T + "]");
    }
    Pairing pairing = pubkey.getPairing();
    Element g = pubkey.getG();
    Element h = pubkey.getH();
    BigInteger r = pairing.getZr().newRandomElement()
                .toBigInteger();
    return g.pow(BigInteger.valueOf(m)).mul(h.pow(r))
                .getImmutable();
}

/**
 *
 * @Title: decrypt
 * @Description: This function is to decrypt the ciphertext
 *                with the public key and the private key.
 * @param c
 *              The ciphertext.
 * @param pubkey
 *              The public key of BGN PKE.
 * @param prikey
 *              The private key of BGN PKE.
 * @return int The plaintext.
 * @throws Exception
 *               If the plaintext is not in [0,1,2,...,n],
 *               there is an exception.
 */
public static int decrypt(Element c, PublicKey pubkey,
        PrivateKey prikey) throws Exception {
    BigInteger p = prikey.getP();
    Element g = pubkey.getG();
    Element cp = c.pow(p).getImmutable();
    Element gp = g.pow(p).getImmutable();
    for (int i = 0; i <= T; i++) {
        if (gp.pow(BigInteger.valueOf(i)).isEqual(cp)) {
            return i;
        }
    }
```

(continued)

```
        throw new Exception(
                "BGN.decrypt(Element c, PublicKey pubkey,
                 PrivateKey prikey): "
                + "plaintext m is not in [0,1,2,...,"
                        + T + "]");
    }

    public static int decrypt_mul2(Element c, PublicKey pubkey,
            PrivateKey prikey) throws Exception {
        BigInteger p = prikey.getP();
        Element g = pubkey.getG();
        Element cp = c.pow(p).getImmutable();
        Element egg = pubkey.getPairing().pairing(g, g).pow(p)
        .getImmutable();
        for (int i = 0; i <= T; i++) {
            if (egg.pow(BigInteger.valueOf(i)).isEqual(cp)) {
                return i;
            }
        }
        throw new Exception(
                "BGN.decrypt(Element c, PublicKey pubkey,
                 PrivateKey prikey): "
                + "plaintext m is not in [0,1,2,...,"
                        + T + "]");
    }

    /**
     * @Title: add
     * @Description: The function supports the homomorphic
     * addition with two ciphertext.
     * @param c1
     *           The ciphertext.
     * @param c2
     *           The ciphertext.
     * @param pubkey
     *           The public key of BGN PKE.
     * @return Element The return value is c1*c2.
     */
    public static Element add(Element c1, Element c2) {
        return c1.mul(c2).getImmutable();
    }

    /**
     * @Title: mul1
     * @Description: The function supports the homomorphic
     *              multiplication with one ciphertext
     *              and one plaintext.
     * @param c
     *           The ciphertext.
     * @param m
```

(continued)

```
 *              The plaintext.
 * @param pubkey
 *              The public key of BNG PKE.
 * @return Element The return value is c^m.
 */
public static Element mul1(Element c1, int m2) {
    return c1.pow(BigInteger.valueOf(m2)).getImmutable();
}

/**
 * @Title: mul2
 * @Description: TODO
 * @param c1
 *              The ciphertext.
 * @param c2
 *              The ciphertext.
 * @param pubkey
 *              The public key of BNG PKE.
 * @return Element The return value is e(c1,c2).
 */
public static Element mul2(Element c1, Element c2,
    PublicKey pubkey) {
    Pairing pairing = pubkey.getPairing();
    return pairing.pairing(c1, c2).getImmutable();
}

/**
 * @Title: selfBlind
 * @Description: The function supports the homomorphic
 *                 self-blinding with one ciphertext
 *                 and one random number.
 * @param c
 *              The ciphertext.
 * @param r
 *              A random number in Z_n.
 * @param pubkey
 *              The public key of BNG PKE.
 * @return Element The return value is c1*h^r2.
 */
public static Element selfBlind(Element c1, BigInteger r2,
    PublicKey pubkey) {
    Element h = pubkey.getH();
    return c1.mul(h.pow(r2)).getImmutable();
}

public static void main(String[] args) {
    BGN bgn = new BGN();
    // Key Generation
    bgn.keyGeneration(512);
    BGN.PublicKey pubkey = bgn.getPubkey();
```

(continued)

```
        BGN.PrivateKey prikey = bgn.getPrikey();

        // Encryption and Decryption
        int m = 5;
        Element c = null;
        int decrypted_m = 0;
        try {
            c = BGN.encrypt(m, pubkey);
            decrypted_m = BGN.decrypt(c, pubkey, prikey);
        } catch (Exception e) {
            e.printStackTrace();
        }
        if (decrypted_m == m) {
            System.out.println("Encryption and Decryption "
                    + "test successfully.");
        }

        // Homomorphic Properties

        // Addition
        int m1 = 5;
        int m2 = 6;
        try {
            Element c1 = BGN.encrypt(m1, pubkey);
            Element c2 = BGN.encrypt(m2, pubkey);
            Element c1mulc2 = BGN.add(c1, c2);
            int decrypted_c1mulc2 = BGN.decrypt(c1mulc2,
            pubkey, prikey);
            if (decrypted_c1mulc2 == (m1 + m2)) {
                System.out.println("Homomorphic addition "
                        + "tests successfully.");
            }
        } catch (Exception e) {
            e.printStackTrace();
        }

        // multiplication-1
        m1 = 5;
        m2 = 6;
        try {
            Element c1 = BGN.encrypt(m1, pubkey);
            Element c1expm2 = BGN.mul1(c1, m2);
            int decrypted_c1expm2 = BGN.decrypt(c1expm2,
            pubkey, prikey);
            if (decrypted_c1expm2 == (m1 * m2)) {
                System.out.println("Homomorphic
                        multiplication-1 "
                        + "tests successfully.");
            }
        } catch (Exception e) {
```

(continued)

```
            e.printStackTrace();
        }

        // multiplication-2
        m1 = 5;
        m2 = 6;
        try {
            Element c1 = BGN.encrypt(m1, pubkey);
            Element c2 = BGN.encrypt(m2, pubkey);
            Element c1pairingc2 = pubkey.getPairing()
                    .pairing(c1, c2).getImmutable();
            int decrypted_c1pairingc2 =
            BGN.decrypt_mul2(c1pairingc2, pubkey, prikey);
            if (decrypted_c1pairingc2 == (m1 * m2)) {
                System.out.println("Homomorphic
                        multiplication-2 "
                        + "tests successfully.");
            }
        } catch (Exception e) {
            e.printStackTrace();
        }

        // self-Blinding
        m1 = 5;
        try {
            BigInteger r2 = pubkey.getPairing().getZr()
                    .newRandomElement().toBigInteger();
            Element c1 = BGN.encrypt(m1, pubkey);
            Element c1_selfblind = BGN.selfBlind(c1,
            r2, pubkey);
            int decrypted_c1_selfblind =
                  BGN.decrypt(c1_selfblind, pubkey, prikey);
            if (decrypted_c1_selfblind == m1) {
                System.out.println("Homomorphic self-blinding "
                        + "tests successfully.");
            }
        } catch (Exception e) {
            e.printStackTrace();
        }
    }
}
```

References

1. S. Goldwasser and S. Micali, "Probabilistic encryption," *J. Comput. Syst. Sci.*, vol. 28, no. 2, pp. 270–299, 1984. [Online]. Available: http://dx.doi.org/10.1016/0022-0000(84)90070-9
2. P. Paillier, "Public-key cryptosystems based on composite degree residuosity classes," in *Advances in Cryptology - EUROCRYPT '99, International Conference on the Theory and Application of Cryptographic Techniques, Prague, Czech Republic, May 2–6, 1999, Proceeding*, 1999, pp. 223–238. [Online]. Available: http://dx.doi.org/10.1007/3-540-48910-X_16
3. T. Okamoto and S. Uchiyama, "A new public-key cryptosystem as secure as factoring," in *Advances in Cryptology - EUROCRYPT '98, International Conference on the Theory and Application of Cryptographic Techniques, Espoo, Finland, May 31 - June 4, 1998, Proceeding*, 1998, pp. 308–318. [Online]. Available: http://dx.doi.org/10.1007/BFb0054135
4. D. Naccache and J. Stern, "A new public key cryptosystem based on higher residues," in *CCS '98, Proceedings of the 5th ACM Conference on Computer and Communications Security, San Francisco, CA, USA, November 3–5, 1998.*, 1998, pp. 59–66. [Online]. Available: http://doi.acm.org/10.1145/288090.288106
5. I. Damgård and M. Jurik, "A generalisation, a simplification and some applications of paillier's probabilistic public-key system," in *Public Key Cryptography, 4th International Workshop on Practice and Theory in Public Key Cryptography, PKC 2001, Cheju Island, Korea, February 13–15, 2001, Proceedings*, 2001, pp. 119–136. [Online]. Available: http://dx.doi.org/10.1007/3-540-44586-2_9
6. D. Boneh, E. Goh, and K. Nissim, "Evaluating 2-dnf formulas on ciphertexts," in *Theory of Cryptography, Second Theory of Cryptography Conference, TCC 2005, Cambridge, MA, USA, February 10–12, 2005, Proceedings*, 2005, pp. 325–341. [Online]. Available: http://dx.doi.org/10.1007/978-3-540-30576-7_18
7. Y. Ishai and A. Paskin, "Evaluating branching programs on encrypted data," in *Theory of Cryptography, 4th Theory of Cryptography Conference, TCC 2007, Amsterdam, The Netherlands, February 21–24, 2007, Proceedings*, 2007, pp. 575–594. [Online]. Available: http://dx.doi.org/10.1007/978-3-540-70936-7_31
8. R. L. Rivest, A. Shamir, and L. M. Adleman, "A method for obtaining digital signatures and public-key cryptosystems," *Commun. ACM*, vol. 21, no. 2, pp. 120–126, 1978. [Online]. Available: http://doi.acm.org/10.1145/359340.359342
9. T. E. Gamal, "A public key cryptosystem and a signature scheme based on discrete logarithms," *IEEE Transactions on Information Theory*, vol. 31, no. 4, pp. 469–472, 1985. [Online]. Available: http://dx.doi.org/10.1109/TIT.1985.1057074
10. N. Koblitz, *A course in number theory and cryptography*. Springer Science & Business Media, 1994, vol. 114.
11. D. Boneh and M. K. Franklin, "Identity-based encryption from the weil pairing," in *Advances in Cryptology - CRYPTO 2001, 21st Annual International Cryptology Conference, Santa Barbara, California, USA, August 19-23, 2001, Proceedings*, 2001, pp. 213–229. [Online]. Available: http://dx.doi.org/10.1007/3-540-44647-8_13
12. D. Boneh, B. Lynn, and H. Shacham, "Short signatures from the weil pairing," in *Advances in Cryptology - ASIACRYPT 2001, 7th International Conference on the Theory and Application of Cryptology and Information Security, Gold Coast, Australia, December 9–13, 2001, Proceedings*, 2001, pp. 514–532. [Online]. Available: http://dx.doi.org/10.1007/3-540-45682-1_30
13. D. Boneh and B. Waters, "Conjunctive, subset, and range queries on encrypted data," in *Theory of Cryptography, 4th Theory of Cryptography Conference, TCC 2007, Amsterdam, The Netherlands, February 21–24, 2007, Proceedings*, 2007, pp. 535–554. [Online]. Available: http://dx.doi.org/10.1007/978-3-540-70936-7_29
14. A. J. Menezes, P. C. Van Oorschot, and S. A. Vanstone, *Handbook of applied cryptography*. CRC press, 1997.

15. C. Gentry, "Fully homomorphic encryption using ideal lattices," in *Proceedings of the 41st Annual ACM Symposium on Theory of Computing, STOC 2009, Bethesda, MD, USA, May 31 - June 2, 2009*, 2009, pp. 169–178. [Online]. Available: http://doi.acm.org/10.1145/1536414.1536440
16. M. van Dijk, C. Gentry, S. Halevi, and V. Vaikuntanathan, "Fully homomorphic encryption over the integers," in *Advances in Cryptology - EUROCRYPT 2010, 29th Annual International Conference on the Theory and Applications of Cryptographic Techniques, French Riviera, May 30 - June 3, 2010. Proceedings*, 2010, pp. 24–43. [Online]. Available: http://dx.doi.org/10.1007/978-3-642-13190-5_2
17. Z. Brakerski, C. Gentry, and V. Vaikuntanathan, "(leveled) fully homomorphic encryption without bootstrapping," in *Innovations in Theoretical Computer Science 2012, Cambridge, MA, USA, January 8–10, 2012*, 2012, pp. 309–325. [Online]. Available: http://doi.acm.org/10.1145/2090236.2090262
18. Z. Brakerski and V. Vaikuntanathan, "Efficient fully homomorphic encryption from (standard) LWE," in *IEEE 52nd Annual Symposium on Foundations of Computer Science, FOCS 2011, Palm Springs, CA, USA, October 22–25, 2011*, 2011, pp. 97–106. [Online]. Available: http://dx.doi.org/10.1109/FOCS.2011.12
19. Z. Brakerski, "Fully homomorphic encryption without modulus switching from classical gapsvp," in *Advances in Cryptology - CRYPTO 2012 - 32nd Annual Cryptology Conference, Santa Barbara, CA, USA, August 19–23, 2012. Proceedings*, 2012, pp. 868–886. [Online]. Available: http://dx.doi.org/10.1007/978-3-642-32009-5_50
20. A. López-Alt, E. Tromer, and V. Vaikuntanathan, "On-the-fly multiparty computation on the cloud via multikey fully homomorphic encryption," in *Proceedings of the 44th Symposium on Theory of Computing Conference, STOC 2012, New York, NY, USA, May 19–22, 2012*, 2012, pp. 1219–1234. [Online]. Available: http://doi.acm.org/10.1145/2213977.2214086
21. C. Gentry, A. Sahai, and B. Waters, "Homomorphic encryption from learning with errors: Conceptually-simpler, asymptotically-faster, attribute-based," in *Advances in Cryptology - CRYPTO 2013 - 33rd Annual Cryptology Conference, Santa Barbara, CA, USA, August 18–22, 2013. Proceedings, Part I*, 2013, pp. 75–92. [Online]. Available: http://dx.doi.org/10.1007/978-3-642-40041-4_5
22. C. Li, R. Lu, H. Li, L. Chen, and J. Chen, "PDA: a privacy-preserving dual-functional aggregation scheme for smart grid communications," *Security and Communication Networks*, vol. 8, no. 15, pp. 2494–2506, 2015. [Online]. Available: http://dx.doi.org/10.1002/sec.1191

Chapter 3
Privacy-Preserving Multidimensional Data Aggregation

Directly applying the homomorphic encryption techniques in the last chapter into smart grid communications cannot capture the unique features of smart grid communications. In this chapter, based on the unique data characteristics, i.e., nearly real time data collection, small size individual data in smart grid, we introduce a privacy-preserving multidimensional data aggregation (PPMDA) scheme [1] for secure smart grid communications.

3.1 Introduction

The concept of smart grid has emerged and been recognized as the next generation of power grid [2–6]. Traditional grid is featured with centralized one-way transmission (from generation plants to customers) and demand-driven response. Smart grid combines traditional grid and information and control technologies. It allows decentralized two-way transmission and reliability- and efficiency- driven response, and aims to provide improved reliability (e.g., self-healing, self-activating, automated outage management), efficiency (e.g., cost-effective power generation, transmission and distribution), sustainability (e.g., accommodation of future alternative and renewable power sources), consumer involvement, and security (physical and cyber).

Smart meters are important components of smart grid. They are two-way communication devices deployed at consumers premise, records power consumption periodically. With smart meters, smart grid is able to collect real-time information about grid operations and status at a control center, through a reliable communications network deployed in parallel to the power transmission and distribution grid, as shown in Fig. 3.1. The control center may be implemented in a distributed way and span different geographic regions. It is responsible for dynamically adjusting power supply to meet demand, and detecting and responding to weaknesses or

R. Lu, *Privacy-Enhancing Aggregation Techniques for Smart Grid Communications*, Wireless Networks, DOI 10.1007/978-3-319-32899-7_3

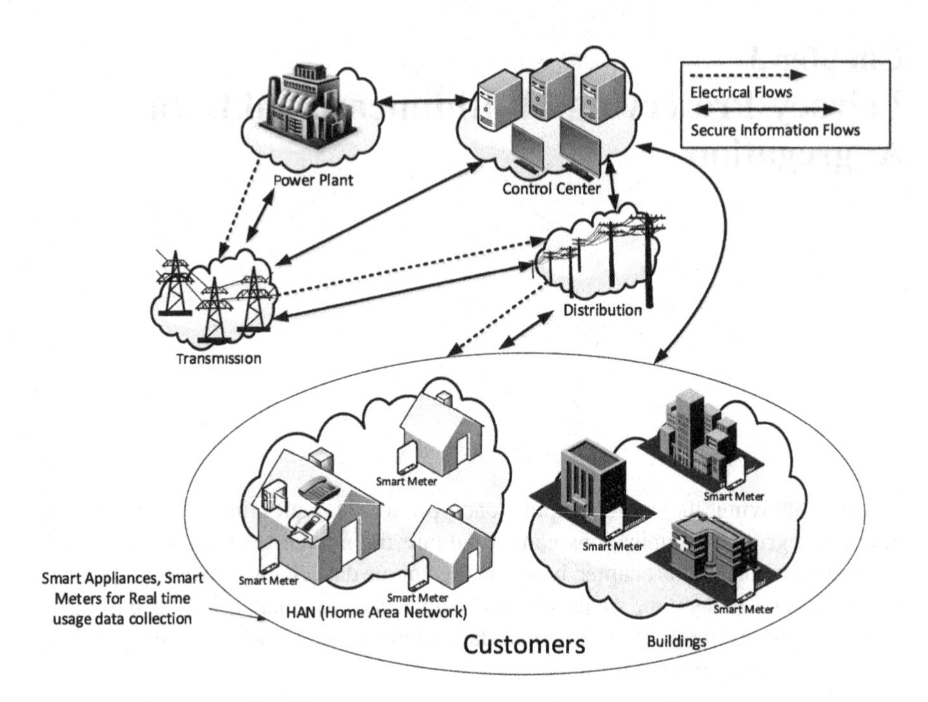

Fig. 3.1 The conceptual architecture of smart grid

failures in the power system in real time. Smart grid also automates reliable power distribution by engaging and empowering customers in utility management. It exposes customers' detailed real-time electricity use information (through smart meters) to utility companies, which may then change electricity price accordingly or even adjust customers' usage by pre-installed load control switches in order to help flatten demand peaks. Customers are allowed to access their own real-time use information through smart grid services. In order to lower their own energy costs and enjoy uninterrupted activities, they will be willing to use energy-efficient appliances and tend to shift power use from peak times to non-peak times.

Cyber security is of paramount importance in smart grid as communications are deeply involved in its operations [7–12]. All the data transmitted in the grid must be authenticated and secured against malicious modification. Privacy (i.e., data confidentiality) is a primary concern from customers point of view as power use information may reveal their physical activities. For example, unusually low daily power consumption of a household and continuous lack of power use for stove and microwave indicate that the home owners are probably away from their home. Such privacy-sensitive information must be protected from unauthorized access. Data confidentiality can be achieved by simple end-to-end encryption. While hiding communication content and protecting privacy, encryption increases data size, and may cause unacceptable communication overhead when power use information is collected at high frequency. Considering that the control center is concerned only with the overall information in a region, the data of individual consumers in the region can be aggregated at a local gateway and forwarded in a compact form to the control center in order to save communication bandwidth.

To preserve user privacy, local gateways should not be able to access the content of consumers data. To enable them to perform data aggregation, homomorphic encryption techniques [13] may be applied for encrypting consumers data. In this technique, a specific linear algebraic manipulation toward the plaintext is equivalent to another one conducted on the ciphertext. This unique feature allows the local gateway to perform summation and multiplication based aggregation on received consumer data without decrypting them. Existing data aggregation schemes [14–16] regard power use information as one-dimensional information. With smart meters being used, it is however multi-dimensional in nature, for example, including the amount of energy consumed, at what time and for what purpose the consumption was, and so on. Taking into account all the dimensions allows finer-grained control and optimization. When multiple dimensions are present, the existing schemes [14–16] will have to process each dimension separately. We further notice that power usage information is often small in size, smaller than the plain text space of the encryption algorithm used. Each time when it is encrypted, its size will be increased to occupy the entire plain text space. Considering the high data collection frequency, multi-dimensional use information and massive number of consumers, the existing data aggregation schemes generate not only huge communication cost but also impose overwhelming process load on local gateways.

To save communication and computation resources, in this paper, we process all the dimensions of the data as a whole rather than separately, and propose a novel efficient and privacy-preserving multi-dimensional data aggregation (PPMDA) scheme. This PPMDA scheme expresses multi-dimensional power use data in a single-dimensional form and supports privacy-preserving aggregation operations on the reformatted data. As a result, data can be efficiently reported to smart grid control center at a high frequency for real-time monitoring and control. The main contributions of this chapter are two-fold.

- Firstly, inspired by the fact that electricity usage data is small in size and multi-dimensional in nature, we present the novel PPMDA scheme that utilizes the homomorphic Paillier PKE [13] to achieve privacy-preserving multi-dimensional data aggregation and efficient smart grid communications. Compared with traditional one-dimensional aggregation schemes [14–16], it leads to dramatically reduced the computation and communication costs.
- Secondly, we analyze the security strength and privacy-preservation ability of PPMDA. In particular, we apply the provable security technique to formally prove that the smart grid control center's response is semantic secure under the chosen plaintext attack. Through comparative performance analysis, we demonstrate that PPMDA is indeed significantly more efficient than existing one-dimensional aggregation schemes [14–16].

The remainder of this chapter is organized as follows. In Sect. 3.2, we introduce our system model, security requirements and our design goal. Then, we present our PPMDA scheme in Sect. 3.3, followed by its security analysis and performance evaluation in Sects. 3.4 and 3.5, respectively. We also discuss the related work in Sect. 3.6. Finally, we draw our conclusions in Sect. 3.7.

3.2 System Model, Security Requirements and Design Goal

In this section, we formalize the system model, security requirements, and identify our design goals.

3.2.1 System Model

In our system model, we mainly focus on how to report residential users' privacy-preserving electricity usage data to the control center in smart grid communications. Specifically, we consider a typical residential area (RA), which comprises a local gateway (GW) connected with smart grid control center, and a large number of residential users $\mathbb{U} = \{U_1, U_2, \cdots, U_w\}$, as shown in Fig. 3.2. The GW is a powerful fog computing device [17, 18], which mainly performs two functions: aggregation and relaying. The aggregation component is responsible for aggregating residential users' electricity usage data into a compressed one, while the relaying component helps residential users with forwarding data to the control center, i.e., to a trusted operation authority (OA) located at control center, and also helps the OA with relaying the responses back to the residential users in the RA as well. In the process of the aggregation and relaying, the GW will also perform some authentication operations to guarantee the data's authenticity and integrity.

Each user $U_i \in \mathbb{U}$ is equipped with various smart meters (SMs), which form a Home Area Network (HAN), and can electronically record the real-time data about electricity use. These near real-time data will then be reported to the OA every a certain period with the relay of the GW. On receiving the reports from residential users, the OA can get the real-time situational awareness so as to make the electricity use more efficient by either carrying out the dynamic price or directly controlling to reduce consumption during peak periods and shift some demands to off-peak hours.

Communication Model. In the residential area RA, the communication between each user $U_i \in \mathbb{U}$ and the local GW is through relatively inexpensive WiFi technology. In other words, within the WiFi coverage of the GW, each $U_i \in \mathbb{U}$ can directly/indirectly communicate with the GW. On the other hand, since the distance between the residential area and the control center is far away, the communication between the GW and the OA is through either wired links or any other links with high bandwidth and low delay. However, although the communication in smart grid is featured with high bandwidth and low delay, since hundred and thousand of residential users scattered at different residential areas in a region will report their

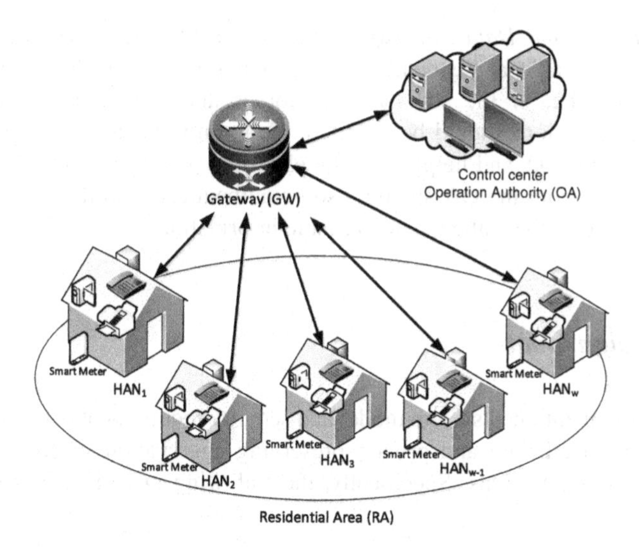

Fig. 3.2 System model under consideration

electricity usage data almost at the same time, the communication efficiency of the GW-to-OA communication is still a challenging issue.

3.2.2 Security Requirements

Security is crucial for the success of secure smart grid communications. In our security model, we consider the OA and the GW are trustable, and the residential users $\mathbb{U} = \{U_1, U_2, \cdots, U_w\}$ are honest as well. However, there exists an adversary $\mathscr{A}$ residing in the RA to eavesdrop the residential users' reports. More seriously, the adversary $\mathscr{A}$ could also intrude in the database of the GW and the smart grid control center to steal the individual user reports. In addition, the adversary $\mathscr{A}$ could also launch some active attacks to threaten the data integrity. Therefore, in order to prevent the adversary $\mathscr{A}$ from learning the users' reports and to detect the adversary $\mathscr{A}$'s malicious actions, the following security requirements should be satisfied in secure smart grid communications.

- *Confidentiality.* Protect individual residential user's reports from the adversary $\mathscr{A}$, i.e., even if $\mathscr{A}$ eavesdrops the WiFi communication in the RA, it cannot identify the contents of the reports; and even if $\mathscr{A}$ steals the data from the control center's and/or the GW's databases, it can also not identify each individual user's data. In such a way, each individual user's electricity usage data can achieve the privacy-preserving requirement. In addition, the confidentiality requirement also includes the OA's responses should be privacy-preserving, i.e., only the legal residential users in the RA can read them.

- *Authentication and Data Integrity.* Authenticating an encrypted report that is really sent by a legal residential user and has not been altered during the transmission, i.e., if the adversary $\mathcal{A}$ forges and/or modifies a report, the malicious operations should be detected. Then, only the correct reports can be received by the OA and helpful for the electricity use monitoring. Meanwhile, the responses from the OA should also be authenticated so that the residential users can receive the authentic and reliable information.

3.2.3 Design Goal

Under the aforementioned system model and security requirements, our design goal is to develop an efficient and privacy-preserving aggregation scheme for secure smart grid communications. Specifically, the following two objectives should be achieved.

- *The security requirements should be guaranteed in the proposed scheme.* As stated above, if the smart grid does not consider the security, the residential users' privacy could be disclosed, and the real-time electricity use reports could be altered. Then, the smart grid cannot step into its flourish. Therefore, the proposed scheme should achieve the confidentiality, authentication and data integrity requirements simultaneously.
- *The communication-effectiveness should be achieved in the proposed scheme.* Although the communication between the OA and the GW is featured with high-bandwidth and low-delay, to support hundred and thousand residential users' reports to the OA at almost the same time, the proposed scheme should also consider the communication-effectiveness, so that the near real-time user reports can be fast transmitted to the OA.

3.3 Proposed PPMDA Scheme

In this section, we propose the efficient and privacy-preserving multi-dimensional data aggregation scheme (PPMDA) for secure smart grid communications, which mainly consists of the following four parts: system initialization, user report generation, privacy-preserving report aggregation, and secure report reading and response.

3.3.1 System Initialization

For a single-authority smart grid system under consideration, it is reasonable to assume a trusted operation authority (OA) can bootstrap the whole system. Specifically, in the system initialization phase, given the security parameters κ, κ_1, OA first generates $(q, P, \mathbb{G}, \mathbb{G}_T, e)$ by running $\mathcal{G}en(\kappa)$, and then calculates the Paillier PKE's public key $(n = p_1 q_1, g)$, and the corresponding private key (λ, μ), where p_1, q_1 are two large primes with $|p_1| = |q_1| = \kappa_1$. Assume that the maximum number of households in a residential area is no more than a constant w, and there are total l types of electricity usage data $(T_1, T_2, \cdots, T_l)$ to be reported in smart grid communications, the value of each type T_i is less than a constant d. Then, OA chooses a super-increasing sequence $\mathbf{a} = (a_1 = 1, a_2, \cdots, a_l)$, where $a_2, \cdots, a_l$ are large primes such that the length $|a_i| \geq \kappa$, $\sum_{j=1}^{i-1} a_j \cdot w \cdot d < a_i$ for $i = 2, \cdots, l$, and $\sum_{i=1}^{l} a_i \cdot w \cdot d < n$. After that, OA computes $(g_1, g_2, \cdots, g_l)$, where

$$g_i = g^{a_i}, \text{ for } i = 1, 2, \cdots, l \tag{3.1}$$

OA also chooses two random elements $Q_1, Q_2 \in \mathbb{G}$, two random numbers $\alpha, x \in \mathbb{Z}_q^*$, and computes $e(P, P)^\alpha$, $Y = xP$. In addition, OA chooses two secure cryptographic hash functions H and H_1, where $H : \{0, 1\}^* \rightarrow \mathbb{G}$ and $H_1 : \{0, 1\}^* \rightarrow \mathbb{Z}_q^*$. In the end, OA publishes the system parameters as

$$\mathsf{pubs} = \left\{ \begin{array}{c} q, P, \mathbb{G}, \mathbb{G}_T, e, n, g_1, \cdots, g_l, \\ Q_1, Q_2, e(P, P)^\alpha, Y, H, H_1 \end{array} \right\} \tag{3.2}$$

and keeps the master keys $(\lambda, \mu, \mathbf{a}, \alpha, x)$ secretly.

When a local gateway (GW) of the residential area (RA) registers itself in the system, it first chooses a random number $x_g \in \mathbb{Z}_q^*$ as the private key, and computes the corresponding public key $Y_g = x_g P$. While when a HAN user $U_i \in \mathbb{U}$ of the RA joins in the system, U_i chooses a random number $x_i \in \mathbb{Z}_q^*$ as the private key, and computes the corresponding public key $Y_i = x_i P$. In addition, the OA uses the master key (α, x) to compute

$$t_{i1} = H_1(U_i || RA || \alpha), t_{i2} = H_1(U_i || RA || x) \tag{3.3}$$

and generates the authorized RA-related key ak_i to U_i, where

$$ak_i = (\alpha P + t_{i1} Y, t_{i1} P, t_{i2} P, t_{i1} Q_1 + t_{i2} Q_2) \tag{3.4}$$

With the authorized key ak_i, U_i can securely receive the response sent by the OA in smart grid communication system.

3.3.2 User Report Generation

In order to achieve the nearly real-time residential users' electricity usage data every η minutes, e.g., $\eta = 15$ min, each HAN user $U_i \in \mathbb{U}$ uses the smart meters to collect l types of data $(d_{i1}, d_{i2}, \cdots, d_{il})$, where each $d_{ij} \leq d$, and performs the following steps:

- *Step-1:* Choose a random number $r_i \in \mathbb{Z}_n^*$, and compute

$$C_i = g_1^{d_{i1}} \cdot g_2^{d_{i2}} \cdot \cdots \cdot g_l^{d_{il}} \cdot r_i^n \bmod n^2 \tag{3.5}$$

- *Step-2:* Use the private key x_i to make a signature σ_i as

$$\sigma_i = x_i H(C_i||RA||U_i||TS) \tag{3.6}$$

  where TS is the current timestamp, which can resist the potential replay attack.
- *Step-3:* Report the encrypted electricity usage data $C_i||RA||U_i||TS||\sigma_i$ to the local gateway GW in the residential area RA.

3.3.3 Privacy-Preserving Report Aggregation

After receiving total w encrypted electricity usage data $C_i||RA||U_i||TS||\sigma_i$, for $i = 1, 2, \cdots, w$, the local GW first checks the timestamp TS and the signature σ_i to verify its validity, i.e., verify whether $e(P, \sigma_i) \stackrel{?}{=} e(Y_i, H(C_i||RA||U_i||TS))$. If it does hold, the signature is accepted, since $e(P, \sigma_i) = e(P, x_i H(C_i||RA||U_i||TS)) = e(Y_i, H(C_i||RA||U_i||TS))$. In order to make the verification efficiently, the GW can perform the batch verification as

$$\begin{aligned} e\left(P, \sum_{i=1}^{w} \sigma_i\right) &= e\left(P, \sum_{i=1}^{w} x_i H(C_i||RA||U_i||TS)\right) \\ &= \prod_{i=1}^{w} e\left(P, x_i H(C_i||RA||U_i||TS)\right) \\ &= \prod_{i=1}^{w} e\left(Y_i, H(C_i||RA||U_i||TS)\right) \end{aligned} \tag{3.7}$$

Then, the time-consuming pairing operations $e(\cdot, \cdot)$ can be reduced from $2w$ to $w+1$ times.

After the validity checking, the GW performs the following steps for privacy-preserving report aggregation:

- *Step-1:* Compute the aggregated and encrypted data C on $C_1, C_2, \cdots, C_w$ as

$$C = \prod_{i=1}^{w} C_i \bmod n^2 \tag{3.8}$$

- *Step-2:* Use the private key x_g to make a signature σ_g as

$$\sigma_g = x_g H(C||RA||GW||TS) \tag{3.9}$$

where TS is the current timestamp.
- *Step-3:* Report the aggregated and encrypted data $C||RA||GW||TS||\sigma_g$ to the operation authority OA.

3.3.4 Secure Report Reading and Response

Upon receiving $C||RA||GW||TS||\sigma_g$, the OA first verifies the validity by checking the signature σ_g on $C||RA||GW||TS$ with $e(P, \sigma_g) = e(Y_g, H(C||RA||GW||TS))$, and then performs the following steps to read the aggregated and encrypted report C, where C is implicitly formed by

$$\begin{aligned} C &= \prod_{i=1}^{w} C_i \bmod n^2 = \prod_{i=1}^{w} g_1^{d_{i1}} \cdot g_2^{d_{i2}} \cdot \cdots \cdot g_l^{d_{il}} \cdot r_i^n \bmod n^2 \\ &= g_1^{\sum_{i=1}^{w} d_{i1}} \cdot g_2^{\sum_{i=1}^{w} d_{i2}} \cdots g_l^{\sum_{i=1}^{w} d_{il}} \cdot \left(\prod_{i=1}^{w} r_i\right)^n \bmod n^2 \\ &= g^{a_1 \sum_{i=1}^{w} d_{i1}} \cdot g^{a_2 \sum_{i=1}^{w} d_{i2}} \cdots g^{a_l \sum_{i=1}^{w} d_{il}} \cdot \left(\prod_{i=1}^{w} r_i\right)^n \bmod n^2 \\ &= g^{a_1 \sum_{i=1}^{w} d_{i1} + a_2 \sum_{i=1}^{w} d_{i2} + \cdots + a_l \sum_{i=1}^{w} d_{il}} \cdot \left(\prod_{i=1}^{w} r_i\right)^n \bmod n^2 \end{aligned} \tag{3.10}$$

- *Step-1:* By taking $M = a_1 \sum_{i=1}^{w} d_{i1} + a_2 \sum_{i=1}^{w} d_{i2} + \cdots + a_l \sum_{i=1}^{w} d_{il}$ and $R = \prod_{i=1}^{w} r_i$, the report $C = g^M \cdot R^n \bmod n^2$ is still a ciphertext of Paillier PKE. Therefore, the OA can use the master key (λ, μ) to recover M as

$$M = a_1 \sum_{i=1}^{w} d_{i1} + a_2 \sum_{i=1}^{w} d_{i2} + \cdots + a_l \sum_{i=1}^{w} d_{il} \bmod n \tag{3.11}$$

Algorithm 1 Recover the aggregated report

1: **procedure** RECOVER THE AGGREGATED REPORT
   Input: $\mathbf{a} = (a_1 = 1, a_2, \cdots, a_l)$ and M
   Output: $(D_1, D_2, \cdots, D_l)$
2: Set $X_l = M$
3: **for** $j = l$ to 2 **do**
4: $X_{j-1} = X_j \bmod a_j$
5: $D_j = \frac{X_j - X_{j-1}}{a_j} = \sum_{i=1}^{w} d_{ij}$
6: **end for**
7: $D_1 = X_1 = \sum_{i=1}^{w} d_{i1}$
8: **return** $(D_1, D_2, \cdots, D_l)$
9: **end procedure**

- *Step-2:* By invoking the Algorithm 1, the OA can recover and store the aggregated data $(D_1, D_2, \cdots, D_l)$, where each $D_j = \sum_{i=1}^{w} d_{ij}$.

The Correctness of Algorithm 1. In Algorithm 1, $X_l = M$, i.e., $X_l = a_1 \sum_{i=1}^{w} d_{i1} + a_2 \sum_{i=1}^{w} d_{i2} + \cdots + a_{l-1} \sum_{i=1}^{w} d_{i(l-1)} + a_l \sum_{i=1}^{w} d_{il} \bmod n$. Since any type of data is less than a constant d, we have

$$
\begin{aligned}
& a_1 \sum_{i=1}^{w} d_{i1} + a_2 \sum_{i=1}^{w} d_{i2} + \cdots + a_{l-1} \sum_{i=1}^{w} d_{i(l-1)} \\
< & a_1 \sum_{i=1}^{w} d + a_2 \sum_{i=1}^{w} d + \cdots + a_{l-1} \sum_{i=1}^{w} d \\
= & \sum_{j=1}^{l-1} a_j w d < a_l
\end{aligned}
\tag{3.12}
$$

Therefore, $X_{l-1} = X_l \bmod a_l = a_1 \sum_{i=1}^{w} d_{i1} + a_2 \sum_{i=1}^{w} d_{i2} + \cdots + a_{l-1} \sum_{i=1}^{w} d_{i(l-1)}$, and

$$
\frac{X_l - X_{l-1}}{a_l} = \frac{a_l \sum_{i=1}^{w} d_{il}}{a_l} = \sum_{i=1}^{w} d_{il} = D_l \tag{3.13}
$$

With the similar procedure, we can also prove each $D_j = \sum_{i=1}^{w} d_{ij}$, for $j = 1, 2, \cdots, l-1$. As a result, the correctness of Algorithm 1 is shown. ■

After analyzing the near real-time electricity usage data $(D_1, D_2, \cdots, D_l)$, the OA responds a message $m \in \mathbb{G}_T$ to inform HAN users $\mathbb{U} = \{U_1, U_2, \cdots, U_w\}$ in the residential area RA about electricity use and control their cost. The concrete steps are performed as follows.

- *Step-1:* The OA first chooses a random number $s \in \mathbb{Z}_q^*$, and computes $\overline{C} = (\overline{C}_1, \overline{C}_2, \overline{C}_3, \overline{C}_4)$, where

$$\begin{cases} \overline{C}_1 = m \cdot e(P,P)^{\alpha s} \bmod q \\ \overline{C}_2 = sP, \overline{C}_3 = sY - sQ_1, \overline{C}_4 = -sQ_2 \end{cases} \tag{3.14}$$

Then, the OA makes the signature $\sigma = xH(\overline{C}||RA||OA||TS)$, where TS is the current timestamp, and sends back $\overline{C}||\sigma$ to the local GW at the residential area RA.

- *Step-2:* Upon receiving $\overline{C}||\sigma$, the GW verifies the validity of $\overline{C}$ by checking whether $e(P,\sigma) = e(Y, H(\overline{C}||RA||OA||TS))$. If it does hold, the GW broadcasts $\overline{C}$ in the residential area RA.
- *Step-3:* After receiving the authenticated $\overline{C}$ from the GW, each HAN user $U_i \in \mathbb{U}$ uses the authorized key $ak_i = (\alpha P + t_{i1}Y, t_{i1}P, t_{i2}P, t_{i1}Q_1 + t_{i2}Q_2)$ to recover m from $\overline{C}$ in the followings.

$$\begin{aligned} &\frac{e(\overline{C}_2, \alpha P + t_{i1}Y)}{e(t_{i1}P, \overline{C}_3)e(t_{i2}P, \overline{C}_4)e(t_{i1}Q_1 + t_{i2}Q_2, \overline{C}_2)} \\ =&\frac{e(sP, \alpha P + t_{i1}Y)}{e(t_{i1}P, sY - sQ_1)e(t_{i2}P, -sQ_2)e(t_{i1}Q_1 + t_{i2}Q_2, sP)} \\ =&\frac{e(sP, \alpha P + t_{i1}Y)}{e(t_{i1}P, sY)} = e(P,P)^{\alpha s} \end{aligned} \tag{3.15}$$

$$\frac{\overline{C}_1}{e(P,P)^{\alpha s}} = \frac{m \cdot e(P,P)^{\alpha s}}{e(P,P)^{\alpha s}} = m \tag{3.16}$$

With the recovered information m, U_i can determine to shift power use from peak times to non-peak times for electricity use efficiency.

3.4 Security Analysis

In this section, we analyze the security properties of the proposed PPMDA scheme. In particular, following the security requirements discussed earlier, our analysis will focus on how the proposed PPMDA scheme can achieve the individual residential user's report privacy preservation, the report's source authentication and data integrity, and the confidentiality of the OA's response.

- *The individual user's report is privacy-preserving in the proposed PPMDA scheme.* In the proposed PPMDA scheme, user U_i's data $(d_{i1}, d_{i2}, \cdots, d_{il})$ sensed by smart meters are formed as $C_i = g_1^{d_{i1}} \cdot g_2^{d_{i2}} \cdot \cdots \cdot g_l^{d_{il}} \cdot r_i^n \bmod n^2$, which can be implicitly expressed as

$$\begin{aligned} C_i &= g_1^{d_{i1}} \cdot g_2^{d_{i2}} \cdot \cdots \cdot g_l^{d_{il}} \cdot r_i^n \bmod n^2 \\ &= g^{a_1 d_{i1}} \cdot g^{a_2 d_{i2}} \cdot \cdots \cdot g^{a_l d_{il}} \cdot r_i^n \bmod n^2 \\ &= g^{a_1 d_{i1} + a_2 d_{i2} + \cdots + a_l d_{il}} \cdot r_i^n \bmod n^2 \end{aligned} \tag{3.17}$$

Obviously, let m_i be $a_1 d_{i1} + a_2 d_{i2} + \cdots + a_l d_{il}$, then the ciphertext $C_i = g^{m_i} \cdot r_i^n \bmod n^2$ is still a valid ciphertext of Paillier PKE. Since Paillier PKE is semantic secure against the chosen plaintext attack, the data $(d_{i1}, d_{i2}, \cdots, d_{il})$ in m_i is also semantic secure and privacy-preserving. Therefore, even though the adversary $\mathscr{A}$ eavesdrops C_i, he still cannot identify the corresponding contents. After collecting all reports $C_1, C_2, \cdots, C_w$ from the residential users, the GW will not recover each reports, instead, it just computes $C = \prod_{i=1}^{w} C_i \bmod n^2$ to perform report aggregation. Therefore, even if the adversary $\mathscr{A}$ intrudes in the GW's database, he cannot get the individual report $(d_{i1}, d_{i2}, \cdots, d_{il})$ either. Finally, after receiving $C = \prod_{i=1}^{w} C_i \bmod n^2$ from the GW, the OA recovers C as $(D_1, D_2, \cdots, D_l)$, where each $D_j = \sum_{j=1}^{w} d_{ij}$, and stores the entry in the database. However, since each $D_j = \sum_{j=1}^{w} d_{ij}$ is an aggregated result, even if the adversary $\mathscr{A}$ steals the data, he still cannot get the individual user U_i's data $(d_{i1}, d_{i2}, \cdots, d_{il})$. Therefore, from the above three aspects, the individual user's report is privacy-preserving in the proposed PPMDA scheme.

- *The authentication and data integrity of the individual user's report and the aggregated report are achieved in the proposed PPMDA scheme.* In the proposed PPMDA scheme, each individual user's report and the aggregated report are signed by BLS short signature [19]. Since the BLS short signature is provably secure under the CDH problem in the random oracle model [20], the source authentication and data integrity can be guaranteed. As a result, the adversary $\mathscr{A}$'s malicious behaviors in the smart grid communications can be detected in the proposed PPMDA scheme.
- *The confidentiality of the OA's response is also achieved in the proposed PPMDA scheme.* When the OA responds m to the residential users in RA, he encrypts it as $\overline{C} = (\overline{C}_1, \overline{C}_2, \overline{C}_3, \overline{C}_4)$. In order to show the confidentiality of m is satisfied, we use the following theorem to prove that $\overline{C} = (\overline{C}_1, \overline{C}_2, \overline{C}_3, \overline{C}_4)$ is semantic secure against chosen-plaintext attack under the assumption that DBDH problem is hard.

Theorem 3.1. *The ciphertext $\overline{C} = (\overline{C}_1, \overline{C}_2, \overline{C}_3, \overline{C}_4)$ is semantic secure against chosen-plaintext attack under the DBDH assumption.*

Proof. Let $a, b, c \in \mathbb{Z}_q^*$, $\tilde{b} \in \{0, 1\}$. If $\tilde{b} = 0$, set $W = e(P, P)^{abc}$; while if $\tilde{b} = 1$, set W to be a random element in $\mathbb{G}_T$. Given (P, aP, bP, cP, W), the DBDH problem is to guess $\tilde{b}$. Assume that there is an adversary $\mathscr{A}$ which runs in polynomial time and has a non-negligible advantage ε to break the semantic security of $\overline{C} = (\overline{C}_1, \overline{C}_2, \overline{C}_3, \overline{C}_4)$ in the PPMDA scheme, then we can construct another adversary $\mathscr{B}$ which has access to $\mathscr{A}$ and achieves a non-negligible advantage to break the DBDH problem.

First, $\mathscr{B}$ is given a DBDH instantiation (P, aP, bP, cP, W) as input, with $W = e(P,P)^{abc}$ when $\tilde{b} = 0$. $\mathscr{B}$ chooses three random numbers $\alpha', \beta_1, \beta_2 \in \mathbb{Z}_q^*$ and sets the system parameters

$$\begin{cases} Y = xP = \underline{aP} \\ Q_1 = \underline{xP + \beta_1 P = aP + \beta_1 P}, Q_2 = \underline{\beta_2 P} \\ e(P,P)^{\alpha} = \underline{e(aP,bP) \cdot e(P,P)^{\alpha'}} \end{cases} \tag{3.18}$$

which implicitly denotes $\alpha = ab + \alpha'$. Because of the randomness of $\alpha', \beta_1, \beta_2$ and aP, bP, the distributions of $(Y, Q_1, Q_2, e(P,P)^{\alpha})$ are unchanged, then the simulated parameters $(Y, Q_1, Q_2, e(P,P)^{\alpha})$ are indistinguishable from the real environment in the eye of $\mathscr{A}$.

Upon receiving $(Y, Q_1, Q_2, e(P,P)^{\alpha})$, $\mathscr{A}$ chooses two messages m_0 and m_1 in $\mathbb{G}_T$ and returns them to $\mathscr{B}$. At this moment, $\mathscr{B}$ flips a bit $b^* \in \{0,1\}$ and generates a ciphertext $\overline{C} = (\overline{C_1}, \overline{C_2}, \overline{C_3}, \overline{C_4})$, where

$$\begin{cases} \overline{C_1} = \underline{m_b^* \cdot W \cdot e(cP, \alpha' P)} \\ \overline{C_2} = \underline{cP} \\ \overline{C_3} = cY - cQ_1 = caP - c(aP + \beta_1 P) = \underline{\beta_1 cP} \\ \overline{C_4} = -cQ_2 = \underline{-\beta_2 cP} \end{cases} \tag{3.19}$$

In the end, $\mathscr{B}$ sends $\overline{C} = (\overline{C_1}, \overline{C_2}, \overline{C_3}, \overline{C_4})$ to $\mathscr{A}$. After receiving $\overline{C}$, $\mathscr{A}$ returns $\mathscr{B}$ a bit b' as the guess of b^*. $\mathscr{B}$ then guesses $\tilde{b} = 0$ if $b' = b^*$. Obviously, when $\tilde{b} = 0$, i.e., $W = e(P,P)^{abc}$, we have

$$\begin{aligned} \overline{C_1} &= m_{b^*} \cdot W \cdot e(cP, \alpha' P) \\ &= m_{b^*} \cdot e(P,P)^{abc} \cdot e(cP, \alpha' P) \\ &= m_{b^*} \cdot e(P,P)^{abc+\alpha' c} = m_{b^*} \cdot e(P,P)^{\alpha c} \end{aligned} \tag{3.20}$$

Then, $\overline{C_1}$ becomes a valid component of the ciphertext. In this case, $\mathscr{A}$ will guess b^* correctly with the probability $\frac{1}{2} + \varepsilon$. Thus, $\Pr[\mathscr{B} \text{ success} | \tilde{b} = 0] = \frac{1}{2} + \varepsilon$. If $\tilde{b} = 1$, $\overline{C_1} = m_{b^*} \cdot W \cdot e(cP, \alpha' P)$ is independent with b^* due to the randomness of W. Therefore, $\Pr[\mathscr{B} \text{ success} | \tilde{b} = 1] = \frac{1}{2}$. Summarizing the above two cases, we have

$$\Pr[\mathscr{B} \text{ success}] = \frac{1}{2}\left(\frac{1}{2} + \varepsilon\right) + \frac{1}{2} \cdot \frac{1}{2} = \frac{1}{2} + \frac{\varepsilon}{2}$$

Since ε is non-negligible, the above result contradicts with the assumption that DBDH problem is hard. As a result, the ciphertext $\overline{C} = (\overline{C_1}, \overline{C_2}, \overline{C_3}, \overline{C_4})$ is semantic secure under the chosen plaintext attack, i.e., the OA's response also achieves the confidentiality in the proposed PPMDA scheme.

From the above analysis, we can see that the proposed PPMDA scheme is secure and privacy-preserving, and can achieve our security design goal.

3.5 Performance Evaluation

In this section, we evaluate the performance of the proposed PPMDA scheme in terms of the computation complexity of residential user, local GW and OA, and the communication overhead of user-to-GW and GW-to-OA communications.

3.5.1 Computation Complexity

For the proposed PPMDA scheme, when a residential user U_i generates an encrypted electricity usage data $C_i||RA||U_i||TS||\sigma_i$, it requires $l+1$ exponentiation operations in $\mathbb{Z}_{n^2}$ to generate C_i, and 1 multiplication operation in $\mathbb{G}$ for σ_i's generation. After receiving the ciphertext from w users, the local GW first verifies the received data by performing a batch verification which includes $w+1$ pairing operations. In addition, the GW should aggregate the reports from different users and generate a signature on the aggregated data. Since the multiplication in $\mathbb{Z}_{n^2}$ is considered negligible compared to exponentiation and pairing operations, the computational cost of aggregation is negligible, and the generation of the signature only includes one multiplication operation in $\mathbb{G}$. As for the OA, it verifies the aggregated data from the GW with two pairing operations and obtains the data by Paillier decryption which includes one exponentiation operation in $\mathbb{Z}_{n^2}$. The OA further sends a response to the GW and in turn to residential users. The generation of the response m costs OA for four multiplication operations in $\mathbb{G}$ and one exponentiation operation in $\mathbb{G}_T$. In order to deliver the response m to users, the extra computational costs for the GW are two pairing operations. After obtaining the response, the extra computational costs for users are from four pairing operations. Denote the computational costs of an exponentiation operation in $\mathbb{Z}_{n^2}$, a multiplication operation in $\mathbb{G}$, an exponentiation operation in $\mathbb{G}_T$ and a pairing operation by C_e, C_m, C_{et} and C_p, respectively. Then, totally for the user, the GW and the OA, the computational cost will be $(l+1)\cdot C_e + C_m + 4\cdot C_p$, $(w+3)\cdot C_p + C_m$, and $2\cdot C_p + C_e + 4\cdot C_m + C_{et}$ in the proposed PPMDA scheme.

The proposed PPMDA scheme enables a residential user to embed multi-dimensional data into one compressed data. It largely reduces the encryption times for users. In the following, for the comparison with PPMDA, we consider a traditional approach (denoted by TRAD), where each user generates a ciphertext for one dimensional data. Under this setting, for l-dimensional data, a user has to generate l ciphertexts and consumes totally $2\cdot l$ exponentiation operations in $\mathbb{Z}_{n^2}$ for the encryption. Including one multiplication operation in $\mathbb{G}_T$ for the signature, the total computational costs for a user would be $2l\cdot C_e + C_m$. After receiving the reports from w users, the GW takes $w + 1$ pairing operations for verifying the signatures, several negligible-cost multiplication operations in $\mathbb{Z}_{n^2}$ for aggregating the ciphertexts, and one multiplication operation in $\mathbb{G}$ for generating the signature. The GW then forwards l ciphertexts to the OA, where each ciphertext contains the sum of one dimensional data of all users. The verification by OA will consume two pairing operations. The number of decryptions executed by OA is l and thus l exponentiation operations in $\mathbb{Z}_{n^2}$ are spent on the decryption. For the response phase, we assume TRAD works exactly the same as the PPMDA scheme. Therefore, the computational costs of a residential user, the GW, and the OA will be $2l \cdot C_e + C_m + 4 \cdot C_p$, $(w+3) \cdot C_p + C_m$, and $2 \cdot C_p + l \cdot C_e + 4 \cdot C_m + C_{et}$, respectively.

We present the computation complexity comparison of PPMDA and TRAD in Table 3.1. Furthermore, we conduct the experiments with PBC [21] and MIRACL [22] libraries running on a 3.0 GHz-processor 512 MB-memory computing machine to study the operation costs. The experimental results indicate that a single exponentiation operation in $\mathbb{Z}_{n^2}$ (the length $|n^2| = 2048$) almost costs 12.4 ms, a single multiplication operation in $\mathbb{G}$ with 160 bits costs 6.4 ms and the corresponding pairing operation costs 20 ms. With the exact operation costs, we depict the variation of computation costs in terms of l in Fig. 3.3. From the figure, it can be obviously shown that the PPMDA scheme largely reduces the computation complexity for both users and the OA.

Table 3.1 Comparison of computation complexity

	PPMDA	TRAD
User	$(l+1) \cdot C_e + C_m + 4 \cdot C_p$	$2l \cdot C_e + C_m + 4 \cdot C_p$
GW	$(w+3) \cdot C_p + C_m$	$(w+3) \cdot C_p + C_m$
OA	$2 \cdot C_p + C_e + 4 \cdot C_m + C_{et}$	$2 \cdot C_p + l \cdot C_e + 4 \cdot C_m + C_{et}$

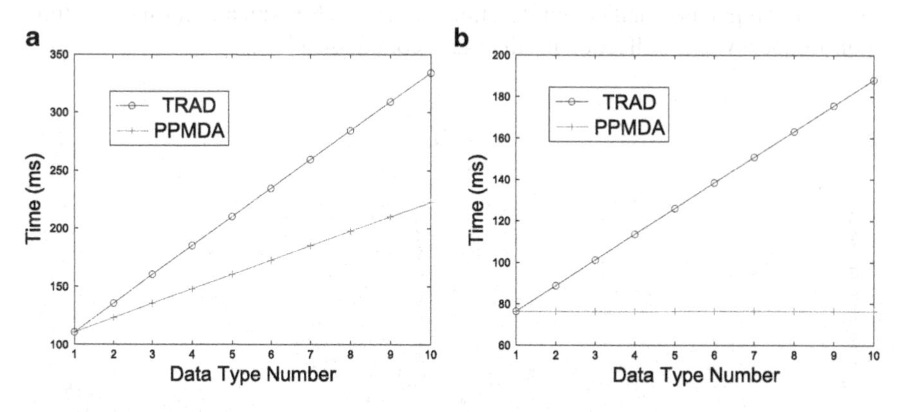

Fig. 3.3 Computation costs of (**a**) each users and (**b**) OA

3.5.2 Communication Overhead

The communications of the proposed PPMDA scheme can be divided into two parts, user-to-GW communication and GW-to-OA communication. We first consider the user-to-GW communication, where users generate their data reports and deliver these reports to the local GW. The data report is in the form of $C_i||RA||U_i||TS||\sigma_i$ for user U_i and its size should be $Sz = 2048 + |RA| + |U_i| + |TS| + 160$ if we choose 1024-bit n and 160-bit $\mathbb{G}$. The GW collects the reports from total w users, indicating that the overall communication overhead between users and the GW is $S_{PPMDA} = w \cdot Sz$. Alternatively, if the traditional TRAD scheme is adopted, each user has to generate a 2048-bit ciphertext for each dimensional data. In this case, the communication overhead of user-to-GW will increase to $S_{trad} = (2048 \cdot l + |U_i| + |RA| + |TS| + 160) \cdot w$. We plot the communication overhead of both schemes in terms of user number w and data types l, as shown in Fig. 3.4, where we set $|RA| + |U_i| + |TS|$ as 100-bit length. It can be seen that the proposed PPMDA scheme always achieves lower communication overhead compared to the TRAD.

Next, we consider the GW-to-OA communication of both PPMDA and TRAD. In PPMDA, the communication is off started by the GW who aims to deliver the aggregated report to the OA. The report is in the form of "$C||RA||GW||TS||\sigma_g$" and with $2048 + |RA| + |GW| + |TS| + 160$ in length. In comparison, in TRAD, different dimensional data has to be aggregated separately, and thus the size of the aggregated reports will be $2048 \cdot l + |RA| + |GW| + |TS| + 160$. In Fig. 3.5, we further plot the communication overhead in terms of user number w and data type l. It is shown that the PPMDA scheme significantly reduces the communication overhead of the GW-to-OA communication.

From the above analysis, the proposed PPMDA scheme is indeed efficient in terms of computation and communication costs, which is suitable for the real-time high-frequency data collection in smart grid communications.

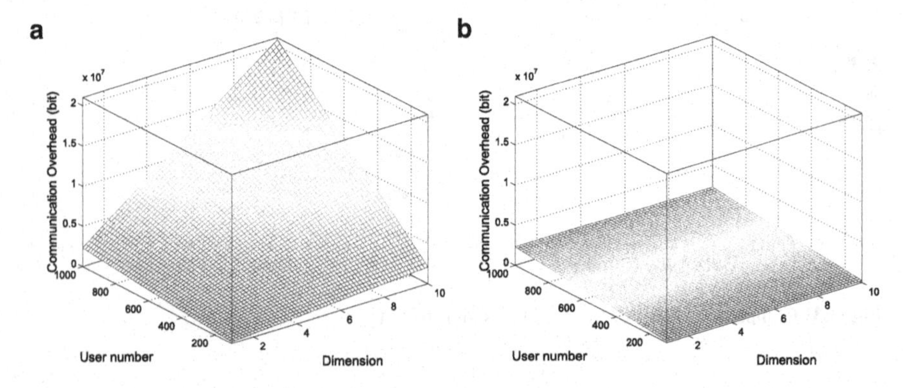

Fig. 3.4 User-to-GW communication overhead. (**a**) TRAD user-to-GW. (**b**) PPMDA user-to-GW

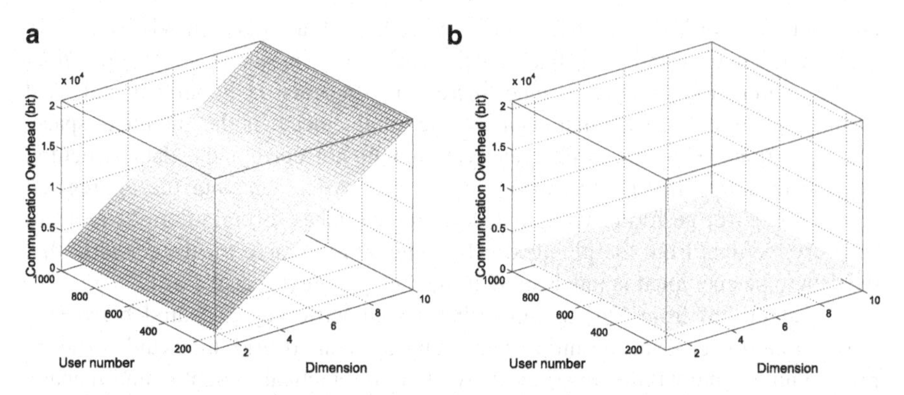

Fig. 3.5 GW-to-OA communication overhead. (**a**) TRAD GW-to-OA. (**b**) PPMDA GW-to-OA

3.6 Related Work

Since large volumes of data from users are to be reported to the OA, it is essential to aggregate individual users' data at intermediate nodes for reducing communication overhead. In most existing secure data aggregation schemes, an intermediate needs to decrypt the received data, aggregate them using aggregation functions, and then encrypt the aggregated result before forwarding it. This process is fairly expensive and risky when intermediate nodes are not trusted. Castelluccia et al. [23] adopted homomorphic encryption techniques to enable efficient aggregation of encrypted data without decryption at intermediate nodes. The proposed scheme is very promising and has triggered considerable followup research [14–16], as discussed below. Castelluccia et al. [14] extended [23] and presented a provable secure and efficient encryption and aggregation scheme. The scheme requires a small number of single-precision additions. It expands packet size only by a small number of bits in the encryption operation and therefore improves computational and communication efficiency. Westhoff et al. [15] indicated that [23] uses different keys per node at the cost of mandatory transmitting of the ID list of the encrypting nodes, resulting in an increased message overhead per monitoring node. They proposed a key pre-distribution scheme that suits the end-to-end encryption of reverse multicast traffic in sensor networks. By using their scheme, a symmetric homomorphic encryption can be applied to increase the efficiency, robustness and flexibility of data aggregation in sensor networks. Shi et al. [16] adopted [23] as a tool, and required users to slice their data into pieces and then to collaboratively aggregate the pieces with others' to preserve user privacy. These previous research works focus on one-dimensional data. How to aggregate multi-dimensional data remains to be a challenging issue.

In smart grid, each individual user has multiple dimensional electricity usage data which is small in size. Performing homomorphic encryption on each dimension of the data requires large computational efforts, and results in large-sized ciphertext and then unaffordable communication cost. Lin et al. [24] introduced super-increasing sequence and perturbation techniques into *compressed data aggregation*. In the aggregation process, a super-increasing sequence is initialized and used to integrate multi-dimensional data as one single piece in the plaintext space. The operations of homomorphic encryption will not corrupt the data structure, i.e., the data in different dimensions are mixed. After receiving the aggregated data, the receiver performs a single decryption and takes several multiplications to recovery the data from the plaintext. This scheme is suitable for the scenario that the data to be encrypted is much smaller than the plaintext space.

The algorithm from [24] assumes that a symmetric key is shared between a sender and a receiver at the initialization phase. In smart grid, thousands of users may communicate with a single gateway. It is not practical to deploy and manage thousands of keys for users and the gateway. In addition, once some residential users stop working and cannot report their data during some periods, the OA cannot use the proper shared keys to recover the correct data since he does not know the participants due to user privacy. As a result, the reliability becomes a big challenge, and this solution cannot directly be applied. In this chapter, we have designed a compressed data aggregation scheme under the public key infrastructure, where users encrypt their reports only with the public key of the OA, and the OA can decrypt the reports only with his private key. Therefore, compared to the symmetric key algorithm in [24], our approach can not only reduce considerable initialization efforts but also achieve high reliability.

3.7 Summary

In this chapter, we have proposed an efficient and privacy-preserving multi-dimensional data aggregation scheme (PPMDA) for secure smart grid communications. It realizes a multi-dimensional data aggregation approach based on the homomorphic Paillier PKE. Compared with the traditional one-dimensional data aggregation methods, PPMDA can significantly reduce computational cost and significantly improve communication efficiency, satisfying the real-time high-frequency data collection requirements in smart grid communications. We have also provided security analysis to demonstrate its security strength and privacy-preserving ability, and performance analysis to show the efficiency improvement.

In PPMDA, OA is trustable and has no interest in knowing each individual user's report. However, if he wishes, he can still know the reports from C_i. In the next chapter, we will introduce a strong privacy-preserving subset data aggregation scheme, where OA also has no idea to know each individual user's report.

References

1. R. Lu, X. Liang, X. Li, X. Lin, and X. Shen, "EPPA: an efficient and privacy-preserving aggregation scheme for secure smart grid communications," *IEEE Trans. Parallel Distrib. Syst.*, vol. 23, no. 9, pp. 1621–1631, 2012. [Online]. Available: http://dx.doi.org/10.1109/TPDS.2012.86
2. F. Li, W. Qiao, H. Sun, H. Wan, J. Wang, Y. Xia, Z. Xu, and P. Zhang, "Smart transmission grid: Vision and framework," *IEEE Transactions on Smart Grid*, vol. 1, no. 1, pp. 168–177, 2010.
3. K. Moslehi and R. Kumar, "A reliability perspective of the smart grid," *IEEE Transactions on Smart Grid*, vol. 1, no. 1, pp. 57–64, 2010.
4. D. Niyato, L. Xiao, and P. Wang, "Machine-to-machine communications for home energy management system in smart grid," *IEEE Communications Magazine*, vol. 49, no. 4, pp. 53–59, 2011.
5. Z. M. Fadlullah, M. M. Fouda, N. Kato, A. Takeuchi, N. Iwasaki, and Y. Nozaki, "Toward intelligent machine-to-machine communications in smart grid," *IEEE Communications Magazine*, vol. 49, no. 4, pp. 60–65, 2011.
6. H. Liang, B. Choi, W. Zhuang, and X. Shen, "Towards optimal energy store-carry-and-deliver for phevs via v2g system," in *Proc. IEEE INFOCOM'12*, Orlando, Florida, USA, March 25–30 2012.
7. P. McDaniel and S. McLaughlin, "Security and privacy challenges in the smart grid," *IEEE Security & Privacy*, vol. 7, no. 3, pp. 75–77, 2009.
8. Z. M. Fadlullah, M. M. Fouda, X. Shen, Y. Nozaki, and N. Kato, "An early warning system against malicious activities for smart grid communications," *IEEE Network Magazine*, to appear.
9. M. M. Fouda, Z. M. Fadlullah, N. Kato, R. Lu, and X. Shen, "A light-weight message authentication scheme for smart grid communications," *IEEE Transactions on Smart Grid*, to appear.
10. M. He and J. Zhang, "A dependency graph approach for fault detection and localization towards secure smart grid," *IEEE Transactions on Smart Grid*, vol. 2, no. 2, pp. 342–351, 2011.
11. Y. Yuan, Z. Li, and K. Ren, "Modeling load redistribution attacks in power systems," *IEEE Transactions on Smart Grid*, vol. 2, no. 2, pp. 382–390, 2011.
12. R. Lu, X. Li, X. Lin, X. Liang, and X. Shen, "GRS: The green, reliability, and security of emerging machine to machine communications," *IEEE Communications Magazine*, vol. 49, no. 4, pp. 28–35, 2011.
13. P. Paillier, "Public-key cryptosystems based on composite degree residuosity classes," in *EUROCRYPT*, 1999, pp. 223–238.
14. C. Castelluccia, A. C.-F. Chan, E. Mykletun, and G. Tsudik, "Efficient and provably secure aggregation of encrypted data in wireless sensor networks," *TOSN*, vol. 5, no. 3, 2009.
15. D. Westhoff, J. Girão, and M. Acharya, "Concealed data aggregation for reverse multicast traffic in sensor networks: Encryption, key distribution, and routing adaptation," *IEEE Transactions on Mobile Computing*, vol. 5, no. 10, pp. 1417–1431, 2006.
16. J. Shi, R. Zhang, Y. Liu, and Y. Zhang, "Prisense: Privacy-preserving data aggregation in people-centric urban sensing systems," in *Infocom*, 2010, pp. 758–766.
17. R. Deng, R. Lu, C. Lai, and T. H. Luan, "Towards power consumption-delay tradeoff by workload allocation in cloud-fog computing," in *2015 IEEE International Conference on Communications, ICC 2015, London, United Kingdom, June 8–12, 2015*, 2015, pp. 3909–3914. [Online]. Available: http://dx.doi.org/10.1109/ICC.2015.7248934
18. T. H. Luan, L. Gao, Z. Li, Y. Xiang, and L. Sun, "Fog computing: Focusing on mobile users at the edge," *CoRR*, vol. abs/1502.01815, 2015. [Online]. Available: http://arxiv.org/abs/1502.01815
19. D. Boneh, B. Lynn, and H. Shacham, "Short signatures from the weil pairing," *Journal of Cryptology*, vol. 17, no. 4, pp. 297–319, 2004.

20. M. Bellare and P. Rogaway, "Random oracles are practical: A paradigm for designing efficient protocols," in *ACM Conference on Computer and Communications Security*, 1993, pp. 62–73.
21. B. Lynn, "PBC library," http://crypto.stanford.edu/pbc/.
22. "Multiprecision integer and rational arithmetic c/c++ library," http://www.shamus.ie/.
23. C. Castelluccia, E. Mykletun, and G. Tsudik, "Efficient aggregation of encrypted data in wireless sensor networks," in *MobiQuitous*, 2005, pp. 109–117.
24. X. Lin, R. Lu, and X. Shen, "MDPA: Multidimensional privacy-preserving aggregation scheme for wireless sensor networks," *Wireless Communications and Mobile Computing (Wiley)*, vol. 10, no. 6, pp. 843–856, 2010.

Chapter 4
Privacy-Preserving Subset Data Aggregation

In order to achieve more accurate data analytics for monitoring and controlling the smart grid, in the last chapter, we discussed a privacy-preserving multi-dimensional data aggregation (PPMDA) scheme. In this chapter, aiming at the same goal, i.e., achieving finer-grained data aggregation, we introduce another flexible privacy-preserving subset data aggregation (PPSDA) scheme [1] for secure smart grid communications.

4.1 Introduction

As the next generation of power grid, smart grid integrates various information and communication technologies (ICT) into power systems to enable the power distribution more reliable and efficient from the power generation, transmission, distribution to end users, as shown in Fig. 4.1. Specifically, due to the two-way communications, smart grid can offer many benefits to both utilities and consumers [2], including (a) overhauling aging equipments in current power system; (b) equipping the power grid to meet increasing demand; (c) decreasing brownouts, blackouts, and surges; (d) giving users to control over their power bills; (e) facilitating real-time troubleshooting; (f) reducing expenses to energy producers; and (g) making renewable power feasible.

Smart meter is one of important components in smart grid communications, which enables residential users to report their nearly real-time electronic consumption data, e.g., every 15 min, to the control center for more reliable monitoring the health of power grid. However, the nearly real time reporting poses a potential threat to user privacy, e.g., personal information at home could be inferred by user's continuously reported data. Therefore, privacy-preserving techniques must be employed in user data reporting for enhancing user's confidence in utilizing smart grid technique.

R. Lu, *Privacy-Enhancing Aggregation Techniques for Smart Grid Communications*, Wireless Networks, DOI 10.1007/978-3-319-32899-7_4

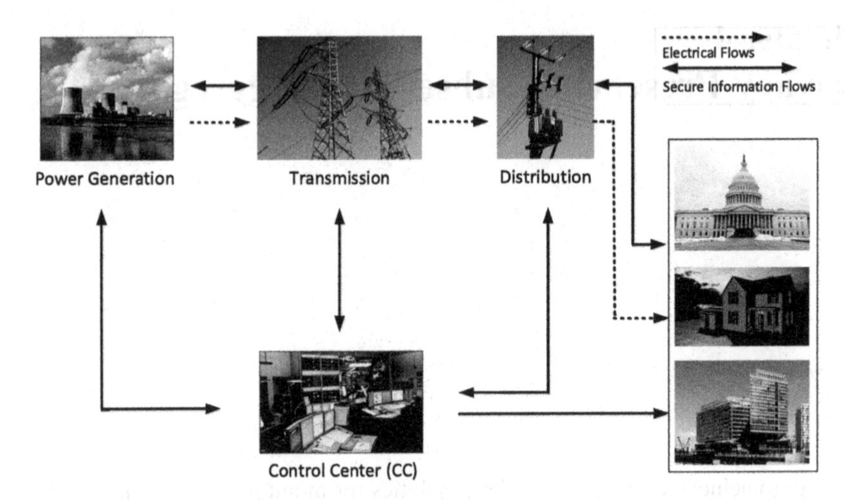

Fig. 4.1 Smart grid—the next generation of power grid

In recent years, to address the above challenge, many privacy-preserving data aggregation schemes have been proposed for smart grid communications [3–11]. However, most of them only support the data aggregation for the whole user set, which sometimes cannot meet the requirements from control center in smart grid communications. For example, the control center needs to know not only the total electronic consumption of the whole set of users, but also the number of users whose electronic consumption is higher than a threshold and the total consumption of these users. Although our previous PPMDA in the last chapter (i.e., EPPA in [11]) can deal with this kind of set aggregation to some extent in smart grid, however, if the control center is assumed as honest-but-curious, each individual user's data can still be obtained by the control center. In this chapter, in order to completely resolve the above problem, we propose a novel privacy-preserving subset data aggregation (PPSDA) scheme for smart grid communications. Specifically, the main contributions of this chapter are three-fold.

- Firstly, by using a group of composite order, we propose a novel privacy-preserving subset data aggregation (PPSDA) scheme. Given a threshold of electronic consumption data, users can be divided into two subsets, then the proposed scheme can use one single aggregated data to aggregate the sum of electronic consumption data in each subset and the corresponding subset size in a privacy-preserving way, which thus supports more accurate data analytics for controlling and monitoring in smart grid.
- Secondly, with formal security proof technique, we show our proposed scheme can achieve each individual user's data privacy preservation.
- Finally, we implement our proposed scheme in Java and run extensive experiments to validate its efficiency in terms of low computational cost and communication overhead.

The remainder of this chapter is organized as follows. In Sect. 4.2, we formalize the system model, security model, and identify our research goal. We present the detailed design of our privacy-preserving set aggregation scheme in Sect. 4.3, followed by the security analysis and performance evaluation in Sects. 4.4 and 4.5, respectively. Section 4.6 reviews some related work and Sect. 4.7 closes the chapter with the summary.

4.2 Models and Design Goal

In this section, we formalize our system model, security model, and identify our design goal on subset data aggregation in smart grid communications.

4.2.1 System Model

In our system model, we focus on the set aggregation at the residential users in smart grid communications. In such a way, our system model mainly includes the following entities: a trusted authority (TA), a control center (CC), a residential gateway (GW) and a set of residential users $\mathbb{U} = \{U_1, U_2, \ldots, U_N\}$, as shown in Fig. 4.2, where N indicates the number of users in the set $\mathbb{U}$, and its maximal value is denoted as $N_{\max}$.

- Trust Authority (TA): TA is a fully trustable entity, whose duty is to manage and distribute key materials to other entities in the system. In general, after key distribution, TA will not be involved in the subsequent data aggregation process.
- Control Center (CC): CC is the core entity in the system, who is responsible for data collecting, processing and analyzing the nearly real-time data from $\mathbb{U}$ for monitoring the health of smart grid.
- Residential Gateway (GW): GW serves as a relay and aggregator role in the system, i.e., GW relays the information from CC to $\mathbb{U}$, and at the same time collects and aggregates the data from $\mathbb{U}$, and forwards the aggregated data to the CC.
- Residential Users $\mathbb{U} = \{U_1, U_2, \ldots, U_N\}$: Each user $U_i \in \mathbb{U}$ is equipped with smart meter, which collects and reports the nearly real-time electricity usage data m_i, e.g., every 15 min, to the CC via the GW.

Different from those previously reported data aggregation schemes [3–11], the subset data aggregation in smart grid communications enables the CC to obtain not only the whole aggregated result $\sum_{U_i \in \mathbb{U}} m_i$ for the set $\mathbb{U}$, but also the partial aggregated result $\sum_{U_j \in \mathbb{U}_1} m_j$ and the size $|\mathbb{U}_1|$ of a subset $\mathbb{U}_1 \subset \mathbb{U}$ *from one single aggregated data*, where $\mathbb{U}_1 = \{U_j | m_j \geq th\}$ and th is a user electronic consumption threshold. With this kind of subset data aggregation, the CC can make more accurate data analytics for monitoring and controlling the smart grid.

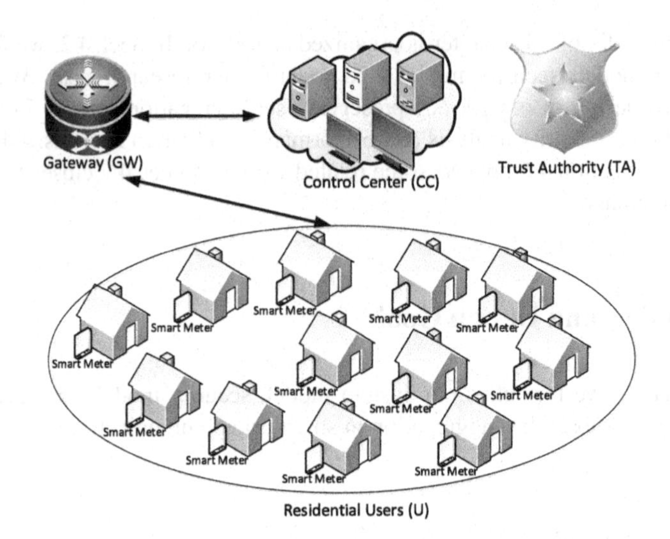

Fig. 4.2 System model under consideration

4.2.2 Security Model

In our security model, we consider both the CC and the GW are *honest-but-curious*. That is, they will faithfully follow the set aggregation protocol, but also attempt to get to know each individual user's data once the condition is satisfied. In addition, residential users $\mathbb{U} = \{U_1, U_2, \ldots, U_N\}$ are also honest, i.e., each U_i won't report false data the CC or collude with the CC to get other users' individual data.

Note that there are possible other attacks, i.e., bad data injection attack [12], DDoS attack, in smart grid communications. Since our focus is on privacy-preserving subset data aggregation, those attacks are currently beyond the scope of this work, and will be discussed in future work.

4.2.3 Design Goal

Our design goal is to develop an efficient and privacy-preserving subset data aggregation scheme for smart grid communications such that the CC can obtain more nearly real-time information from one single aggregated data. Specifically, the following two desirable goals should be satisfied.

- *The proposed scheme should be privacy-preserving.* Only the CC can read the subset aggregation results in the proposed scheme, and no one (including the CC) can read each individual user data.
- *The proposed scheme should be efficient.* Not only the encryption at user side, aggregation at gateway, but also the decryption at control center should be

efficient in terms of computational cost. In addition, the set aggregation, like other data aggregation schemes [3–11], should use one single aggregation data for transmission so as to achieve communication efficiency.

4.3 Proposed PPSDA Scheme

In this section, we propose our privacy-preserving subset data aggregation (PPSDA) scheme, which is mainly comprised of three parts: system initialization, encryption at user side, aggregation at gateway, and decryption at control center. Before plunging into the details, we first review hard problems in group with composite order [13], which serves as the basis of the proposed scheme.

4.3.1 Hard Problems in Group with Composite Order

Let κ be the security parameter and $\mathbb{G}(g, \times)$ be a cyclic multiplication group generated by g with composite order n, where $n = pq$ and p, q are two large prime numbers with $|p| = |q| = \kappa$. Then, the Decisional Diffie-Hellman (DDH) Problem and Subgroup Decision (SGD) Problem in $\mathbb{G}$ are described as follows:

Definition 4.1 (DDH Problem). Given $(g, g^a, g^b, Z) \in \mathbb{G}$ with unknown $a, b \in \mathbb{Z}_n^*$, to determine whether or not $Z = g^{ab} \in \mathbb{G}$. We say that (τ, ϵ)-DDH Assumption holds in $\mathbb{G}$ if no τ-time algorithm has advantage at least ϵ in solving the above DDH problem in $\mathbb{G}$.

Definition 4.2 (SGD Problem). Without knowing the factorization of the group order $n = pq$, given an element $x \in \mathbb{G}$, to decide whether x is an element of subgroup in $\mathbb{G}$ with order p or q. We say that (t, ϵ)-SGD Assumption holds in $\mathbb{G}$ if no t-time algorithm has advantage at least ϵ in solving the above SGD problem in group $\mathbb{G}$.

4.3.2 Description of The Proposed Scheme

4.3.2.1 System Initialization

Given the security parameter κ, a number $N_{\max}$ indicating the maximal user number in $\mathbb{U}$, a small number Δ indicating the maximal electricity consumption data in every t interval time, the control center (CC) first randomly chooses two large primes p, q such that $|p| = |q| = \kappa$, $2q > p$, and $p - q > N_{\max} \cdot \Delta$, computes $n = pq$, and also chooses a cyclic multiplication group $\mathbb{G}$ generated by g with the composite order n. After that, the CC chooses a cryptographic hash function $H : \{0, 1\}^* \rightarrow \mathbb{G}$ and computes $h_0 = g^q$ and $h_1 = g^p$ in $\mathbb{G}$. Finally, the CC keeps $sk = p$ as the private key, and publishes the public key $pk = (n, g, h_0, h_1, H)$.

After verifying the validation of the CC's public key pk, the trusted authority (TA) chooses N random numbers $x_i \in \mathbb{Z}_n^*$, $i = 1, 2, \cdots, N$, and computes $x_0 \in \mathbb{Z}_n^*$ such that

$$x_0 + \sum_{i=1}^{N} x_i = 0 \bmod n \tag{4.1}$$

Finally, the TA sends x_0 as an additional decryption secret key to the CC, and x_i as a secret key to each corresponding user $U_i \in \mathbb{U} = \{U_1, U_2, \cdots, U_N\}$ via secure channels.

4.3.2.2 Encryption at User Side

At every time interval t, e.g., every 15 min, the CC chooses a threshold th and requests subset data aggregation from all residential users $\mathbb{U} = \{U_1, U_2, \cdots, U_N\}$. If a user U_i's electricity consumption data m_i is greater than or equal to the threshold th, i.e., $m_i \geq th$, U_i lies in the subset $\mathbb{U}_1 \subset \mathbb{U}$. Otherwise, U_i lies in the subset $\mathbb{U}_0 \subset \mathbb{U}$. Obviously, $\mathbb{U} = \mathbb{U}_1 \cup \mathbb{U}_0$, $\mathbb{U}_1 \cap \mathbb{U}_0 = \phi$.

Each user $U_i \in \mathbb{U}$ runs the following steps to encrypt his electricity consumption data m_i. Note that, because the time interval t is small, it is reasonable to assume m_i lies in a small set $\{0, 1, 2, \cdots, \Delta\}$.

- *Step 1:* U_i compares his consumption data m_i with the threshold th. If $m_i \geq th$, $U_i \in \mathbb{U}_1$ uses his secret key x_i to compute

$$c_i = g^{m_i} \cdot h_1 \cdot H(t)^{x_i} \tag{4.2}$$

  Otherwise, if $m_i < th$, $U_i \in \mathbb{U}_0$ computes

$$c_i = h_0^{m_i} \cdot H(t)^{x_i} \tag{4.3}$$

- *Step 2:* U_i sends c_i to the gateway (GW).

4.3.2.3 Aggregation at Gateway

After receiving all c_i, $i = 1, 2, \cdots, N$, from the residential users $\mathbb{U}$, the GW performs the following aggregation,

$$
\begin{aligned}
C &= \prod_{i=1}^{N} c_i \\
&= g^{\sum_{U_i \in \mathbb{U}_1} m_i} \cdot h_0^{\sum_{U_j \in \mathbb{U}_0} m_j} \cdot h_1^{\sum_{U_i \in \mathbb{U}_1} 1} \cdot H(t)^{\sum_{i=1}^{N} x_i} \\
&= g^{\sum_{U_i \in \mathbb{U}_1} m_i} \cdot h_0^{\sum_{U_j \in \mathbb{U}_0} m_j} \cdot h_1^{|\mathbb{U}_1|} \cdot H(t)^{\sum_{i=1}^{N} x_i}
\end{aligned} \tag{4.4}
$$

and forwards the result C to the CC.

4.3.2.4 Decryption at Control Center

Upon receiving C, the CC performs the following steps to recover the aggregated data

- *Step 1:* the CC uses his secret key x_0 to compute

$$
\begin{aligned}
D =& C \cdot H(t)^{x_0} \\
=& g^{\sum_{U_i \in \mathbb{U}_1} m_i} \cdot h_0^{\sum_{U_j \in \mathbb{U}_0} m_j} \cdot h_1^{|\mathbb{U}_1|} \cdot H(t)^{\sum_{i=1}^{N} x_i + x_0} \\
& \xrightarrow{\because \sum_{i=1}^{N} x_i + x_0 = 0 \bmod n} \\
=& g^{\sum_{U_i \in \mathbb{U}_1} m_i} \cdot h_0^{\sum_{U_j \in \mathbb{U}_0} m_j} \cdot h_1^{|\mathbb{U}_1|}
\end{aligned} \tag{4.5}
$$

- *Step 2:* Because the CC knows h_0 is an element in the subgroup of $\mathbb{G}$ with order p, the CC uses the private key p to compute

$$
\begin{aligned}
\bar{D} =& D^p \\
=& \left(g^{\sum_{U_i \in \mathbb{U}_1} m_i} \cdot h_0^{\sum_{U_j \in \mathbb{U}_0} m_j} \cdot h_1^{|\mathbb{U}_1|} \right)^p \\
& \xrightarrow{\because h_0{}^p = 1} \\
=& \left(g^{\sum_{U_i \in \mathbb{U}_1} m_i} \cdot h_1^{|\mathbb{U}_1|} \right)^p \\
=& h_1^{\sum_{U_i \in \mathbb{U}_1} m_i + p \cdot |\mathbb{U}_1|}
\end{aligned} \tag{4.6}
$$

  and applies the Algorithm 2 to obtain the values of $\sum_{U_i \in \mathbb{U}_1} m_i$ and $|\mathbb{U}_1|$, where $|\mathbb{U}_1|$ is the size of subset $\mathbb{U}_1$.
- *Step 3:* After calculating $\sum_{U_i \in \mathbb{U}_1} m_i$ and $|\mathbb{U}_1|$ in the last step, the CC now computes

Algorithm 2 Decrypt $\sum_{U_i \in \mathbb{U}_1} m_i$ and $|\mathbb{U}_1|$

```
1: procedure DECRYPTION
2:     on input of D̄ = h_1^{Σ_{U_i∈U_1} m_i + p·|U_1|}
3:     for (i = 0; i <= N; i++) do
4:         for (j = 0; j <= N · Δ; j++) do
5:             if (h_1^j · (h_1^p)^i == D̄) then
6:                 set Σ_{U_i∈U_1} m_i = j, |U_1| = i
7:                 break
8:             end if
9:         end for
10:    end for
11:    return Σ_{U_i∈U_1} m_i, |U_1|
12: end procedure
```

$$\hat{D} = \frac{D}{g^{\sum_{U_i \in \mathbb{U}_1} m_i} \cdot h_1^{|\mathbb{U}_1|}} = h_0^{\sum_{U_j \in \mathbb{U}_0} m_j} \tag{4.7}$$

Because $\sum_{U_j \in \mathbb{U}_0} m_j$ is in the range of $[0, N \cdot \Delta]$, $\sum_{U_j \in \mathbb{U}_0} m_j$ can be efficiently recovered from $\hat{D}$ by using Pollard's lambda method [14].

- *Step 4:* Because $\mathbb{U} = \mathbb{U}_1 \cup \mathbb{U}_0$, $\mathbb{U}_1 \cap \mathbb{U}_0 = \phi$, we can calculate the size of the subset $\mathbb{U}_0$ as $|\mathbb{U}_0| = N - |\mathbb{U}_1|$. Finally, the CC obtains the set aggregation shown in Table 4.1. From the results in Table 4.1, the CC can easily compute the whole aggregation value

$$\sum_{i=1}^{N} m_i = \sum_{U_i \in \mathbb{U}_1} m_i + \sum_{U_j \in \mathbb{U}_0} m_j \tag{4.8}$$

and run other more accurate data analytics algorithms for controlling and monitoring the smart grid.

Correctness. Obviously, the correctness of the proposed scheme depends upon whether the Algorithm 2 can produce a unique solution (x, y) such that $\sum_{U_i \in \mathbb{U}_1} m_i = x$, $|\mathbb{U}_1| = y$, and $0 \le x \le N \cdot \Delta$, $0 \le y \le N$. In the following, we use Theorem 4.1 to show its correctness.

Table 4.1 The results of set aggregation

	Size	Aggregated data
$\mathbb{U}_1$	$\lvert\mathbb{U}_1\rvert$	$\sum_{U_i \in \mathbb{U}_1} m_i$
$\mathbb{U}_0$	$\lvert\mathbb{U}_0\rvert$	$\sum_{U_j \in \mathbb{U}_0} m_j$

Theorem 4.1. *Let* $\hat{D}$ *be* $h_1^{\sum_{U_i \in \mathbb{U}_1} m_i + p \cdot |\mathbb{U}_1|}$ *derived from a valid aggregation ciphertext C with the operations in Eqs. (4.5)–(4.6). Then, there exists a unique solution* (x, y) *such that* $\sum_{U_i \in \mathbb{U}_1} m_i = x$*,* $|\mathbb{U}_1| = y$*, and* $0 \leq x \leq N \cdot \Delta$*,* $0 \leq y \leq N$*.*

Proof. Assume that, by running Algorithm 2, we have two solutions (x, y), (x', y') such that

$$\hat{D} = h_1^{x+p \cdot y} = h_1^{x'+p \cdot y'}$$

where $0 \leq x, x' \leq N \cdot \Delta$ and $0 \leq y, y' \leq N$. Because the order of h_1 is q, i.e., $h_1{}^q = 1$, we have

$$x + py = x' + py' \bmod q$$

Without loss of generality, we assume $y > y'$, then

$$\begin{aligned}
x' =& x + p(y - y') \bmod q \\
=& [(x \bmod q) + (p \bmod q) \cdot ((y - y') \bmod q)] \bmod q \\
& \xrightarrow{\because 2q > p \text{ and } p - q > N_{\max} \cdot \Delta, \therefore (p \bmod q) > N_{\max} \cdot \Delta} \\
>& [(x \bmod q) + N_{\max} \cdot \Delta \cdot ((y - y') \bmod q)] \bmod q \\
=& x + N_{\max} \cdot \Delta \cdot (y - y') \bmod q \\
\geq& x + N \cdot \Delta \cdot (y - y') \bmod q
\end{aligned}$$

which indicates that $x' > N \cdot \Delta$ and contradicts with the constraint $0 \leq x' \leq N \cdot \Delta$. Therefore, there only exists one unique solution (x, y) such that $\sum_{U_i \in \mathbb{U}_1} m_i = x$, $|\mathbb{U}_1| = y$, and $0 \leq x \leq N \cdot \Delta, 0 \leq y \leq N$.

Note that, as the three values of $\sum_{U_i \in \mathbb{U}_1} m_i$, $\sum_{U_j \in \mathbb{U}_0} m_j$, and $|\mathbb{U}_1|$ in $D = g^{\sum_{U_i \in \mathbb{U}_1} m_i} \cdot h_0^{\sum_{U_j \in \mathbb{U}_0} m_j} \cdot h_1^{|\mathbb{U}_1|}$ are small, it is possible to use the brute force method to directly guess a solution (x, y, z) such that $x \in [0, N \cdot \Delta], y \in [0, N \cdot \Delta], z \in [0, N]$ and $D = g^x h_0{}^y h_1{}^z$. However, the complexity is $O(N^3 \cdot \Delta^2)$, while the proposed scheme only requires $O(N^2 \cdot \Delta + \sqrt{N \cdot \Delta})$.

Source Code. The sample java source code of PPSDA is available in Appendix in this chapter.

4.4 Security Analysis

In this section, we analyze the privacy properties of the proposed scheme. In specific, following the security model discussed earlier, we will show that (a) the CC cannot read each individual user's data, and (b) no one, expect for the CC, can read the set aggregation results.

- *The CC cannot read each individual user's data in the proposed scheme.*

First, no matter whether a user U_i is in subset $\mathbb{U}_1$ or $\mathbb{U}_0$, we can always unify U_i's message M_i and the corresponding ciphertext c_i as

$$M_i = g^{a_{i1}} h_0^{a_{i2}} h_1^{a_{i3}}, \quad c_i = g^{a_{i1}} h_0^{a_{i2}} h_1^{a_{i3}} H(t)^{x_i} = M_i H(t)^{x_i}$$

where

$$\begin{cases} a_{i1} = m_i, a_{i2} = 0, a_{i3} = 1, \text{ if } U_i \in \mathbb{U}_1; \\ a_{i1} = 0, a_{i2} = m_i, a_{i3} = 0, \text{ if } U_i \in \mathbb{U}_0. \end{cases}$$

Based on the above decryption procedure at the control center, only if knowing $M_i = g^{a_{i1}} h_0^{a_{i2}} h_1^{a_{i3}}$, the CC can use the private key p to recover (a_{i1}, a_{i2}, a_{i3}). Therefore, in order to keep U_i's data privacy, we need to prove that the CC cannot get M_i from $c_i = M_i H(t)^{x_i}$. To formally show this point, we first assume that U_i's public key $Y_i = g^{x_i}$ corresponding to the secret key x_i is available to the CC. Then, we prove in Theorem 4.2 that, even though the CC obtains Y_i, the CC still cannot know M_i under DDH assumption and in the random oracle model [15].

Theorem 4.2. *Let $\mathcal{A}$ be any chosen-plaintext adversary against the user U_i's ciphertext $c_i = M_i \cdot H(t)^{x_i}$ with time τ. After q_h queries to the random oracles, its advantage is a non-negligible ϵ. Then, the DDH problem in $\mathbb{G}$ can be solved with another probability ϵ' with time τ', where*

$$\epsilon' = \frac{\epsilon}{2}, \quad \tau' \leq \tau + q_h \cdot T_h$$

with T_h denotes the time cost for each hash query.

Proof. We now use sequence games [16] to formally prove the theorem, i.e., showing the ciphertext $c_i = M_i \cdot H(t)^{x_i}$ is indistinguishable against $\mathcal{A}$ under chosen-plaintext attack (IND-CPA). We define a sequence of games $Game_1, Game_2, \cdots$ of modified attacks starting from the actual game $Game_0$. With these incremental games, we reduce a DDH problem instance, i.e., given (g, g^a, g^b, Z) for unknown $a, b \in \mathbb{Z}_n^*$ to determine whether or not $Z = g^{ab}$, to an IND-CPA attack against $c_i = M_i \cdot H(t)^{x_i}$. In other words, we will show that the adversary $\mathcal{A}$ can help to solve the DDH problem in $\mathbb{G}$.

$Game_0$: This is a real game in the random oracle model. We take the role of user U_i, and know the public and private key pair $(Y_i = g^{x_i}, x_i)$. The adversary $\mathcal{A}$ knows the public key $Y_i = g^{x_i}$ and is allowed to access a random oracle $\mathcal{O}_H$. At some time, $\mathcal{A}$ outputs two messages (M_{i0}, M_{i1}) and a time point t^* for encryption query. Then, we toss a coin to get a random $\beta \in \{0, 1\}$, encrypt and return $c_i^* = M_{i\beta} \cdot H(t^*)^{x_i}$ to $\mathcal{A}$. Finally, $\mathcal{A}$ outputs his guess $\beta' \in \{0, 1\}$ on β. We denote $Guess_0$ as the event that $\beta = \beta'$ in $Game_0$ and use the notation $Guess_j$ for the same meaning in any game $Game_j$. Then, based on the definition, we have

$$\Pr[Guess_0] = \Pr[\beta = \beta'], \quad \epsilon = 2\Pr[\beta = \beta'] - 1$$

Game$_1$*:* In this game, we embed the challenge (g, g^a, g^b, Z) into the game, i.e., simulating the random oracle $\mathcal{O}_H$ by maintaining a hash list Λ_H. When a fresh query on time t_i is asked, we first choose a random number $r_i \in \mathbb{Z}_n^*$, set and return $H(t_i) = g^{b \cdot r_i}$ to $\mathscr{A}$, and also store $(t_i, r_i, H(t_i))$ in Λ_H. Because $H(t_i) = g^{b \cdot r_i}$ is uniformly distributed in $\mathbb{G}$, as a result, this game is perfectly indistinguishable from the previous one. Therefore

$$\Pr[Guess_1] = \Pr[Guess_0]$$

Game$_2$*:* In this game, we replace the public key $Y_i = g^{x_i}$ with g^a. Once $Y_i = g^a$, we do not know the corresponding private key a. Therefore, when $\mathscr{A}$ sends (M_{i0}, M_{i1}, t^*) for a request, we perform the following simulation steps.

- Find the entry $(t_x, r_x, H(t_x))$ in Λ_H such that $t_x = t^*$.
- Compute $c_i = M_{ib^*} \cdot Z^{r_x}$ and return c_i to $\mathscr{A}$.

Now, we define the event B that $Z = g^{ab} \in \mathbb{G}$. If the event B really happens, then $c_i = M_{i\beta} \cdot Z^{r_x}$ is a valid ciphertext under the public key $Y_i = g^a$ and $H(t^*) = g^{b \cdot r_x}$. Therefore, the adversary $\mathscr{A}$ can exert his capability to guess whether $\beta = \beta'$ on $c_i = M_{i\beta} \cdot Z^{r_x}$. That is,

$$\Pr[Guess_2|B] = \Pr[Guess_1]$$

However, if the event B doesn't occur, i.e., $Z \neq g^{ab} \in \mathbb{G}$, then $\mathscr{A}$ can only randomly guess whether $\beta = \beta'$ with $1/2$ probability. Thus,

$$\Pr[Guess_2|\bar{B}] = \frac{1}{2}$$

Therefore, from the above analysis, we can solve the DDH problem in $\mathbb{G}$ with the advantage ϵ', where

$$\begin{aligned}\epsilon' &= \Pr[Guess_2|B] - \Pr[Guess_2|\bar{B}] = \Pr[Guess_1] - \frac{1}{2} \\ &= \Pr[Guess_0] - \frac{1}{2} = \Pr[\beta = \beta'] - \frac{1}{2} = \frac{\epsilon}{2}\end{aligned}$$

By a simple computation, we can easily obtain the claimed bound for $\tau' \leq \tau + q_h \cdot T_h$. Thus, the proof is completed. ■

From Theorem 4.2, we can see, even though the adversary $\mathscr{A}$ knows $Y = g^{x_i}$, the ciphertext $c_i = M_i \cdot H(t)^{x_i}$ is still secure. Therefore, we can conclude that the CC cannot read each individual user's data, and identify whether a user in $\mathbb{U}_1$ or $\mathbb{U}_0$ in the proposed scheme.

- *No one, except the CC, can read the set aggregation results in the proposed scheme.*

In the proposed scheme, the aggregation ciphertext C is in the form of

$$C = g^{\sum_{U_i \in \mathbb{U}_1} m_i} \cdot h_0^{\sum_{U_j \in \mathbb{U}_0} m_j} \cdot h_1^{|\mathbb{U}_1|} \cdot H(t)^{\sum_{i=1}^{N} x_i} \in \mathbb{G}$$

Without knowing the secret key x_0 such that $x_0 + \sum_{i=1}^{n} = 0 \mod n$, $H(t)^{\sum_{i=1}^{N} x_i}$ cannot be removed from C. Therefore, only the CC can compute

$$D = C \cdot H(t)^{x_0} = g^{\sum_{U_i \in \mathbb{U}_1} m_i} \cdot h_0^{\sum_{U_j \in \mathbb{U}_0} m_j} \cdot h_1^{|\mathbb{U}_1|}$$

and read the set aggregation results $\sum_{U_i \in \mathbb{U}_1} m_i$, $\sum_{U_j \in \mathbb{U}_0} m_j$, and $|\mathbb{U}_1|$ in the proposed scheme.

From the above analysis, we can conclude that our proposed scheme is a secure and privacy-preserving subset data aggregation scheme for smart grid communications.

4.5 Performance Evaluation

In this section, we evaluate our proposed privacy-preserving set aggregation scheme in terms of computational cost and communications overheads. Specifically, we implement our scheme by Java (JDK 1.8) and run our experiments on a Laptop with 3.1 GHz processor, 8GB RAM, and Window 7 platform. The detailed parameter settings are shown in Table 4.2.

Table 4.2 The parameter settings

Parameter	Value
κ	$\kappa = 512$
$\mathbb{G}$	$\mathbb{G}$ is a subgroup of $\mathbb{Z}_P^*$ of order $n = pq$, where $P = 2pq+1$ is a large prime, and p, q are also two primes with $\lvert p \rvert = \lvert q \rvert = \kappa$
$N_{\max}$	$N_{\max} = 500$
N	$N = 50, 100, 150, 200, 250, 300, 350, 400, 450, 500$
Δ	$\Delta = 10$
th	The threshold th is randomly chosen from $[1, \Delta]$

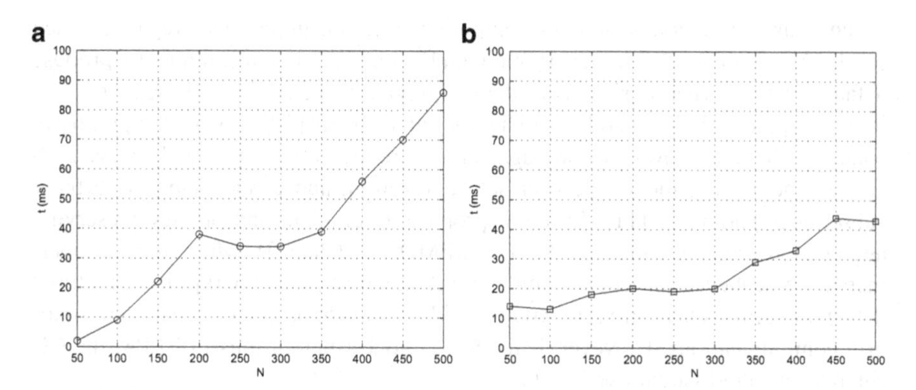

Fig. 4.3 Computational costs of aggregation and decryption varying with N. (**a**) Aggregation at GW. (**b**) Decryption at CC

Although the decryption complexity of the proposed scheme has been $O(N^2 \cdot \Delta + \sqrt{N\Delta})$, reduced from $O(N^3 \cdot \Delta^2)$, and can be acceptable by the powerful control center in smart grid communications, we still establish a hash table (stored in a zip file around 40 M) for accelerating the looking-up process in decryption in our experiment, where each entry in Hash table is the hash value of $h_1^j \cdot (h_1^p)^i$, with $0 \leq j \leq N_{\max} \cdot \Delta$ and $0 \leq i \leq N_{\max}$. We run our experiments 10 times, and the average results are reported below.

Computational Cost. No matter whether a user belongs to subset $\mathbb{U}_1$ or $\mathbb{U}_0$, the average encryption at user side only takes 3.46 ms, which is extremely efficient. Figure 4.3 shows the computational costs of aggregation at GW and decryption at CC varies with the number of user N from 50 to 500 with the increment of 50. From the figure, we can see both of them are efficient, and the number of users N has a little effect on the aggregation and decryption, after a hash table used for looking-up in decryption is established in advance.

Communication Cost. When the length of p, q is $|p| = |q| = 512$, the length of $P = 2pq + 1$ is 1025 bits. Thus, any ciphertext (including c_i and C) in the subgroup $\mathbb{G}$ of $\mathbb{Z}_P^*$ is less than or equal to 1025 bits.

4.6 Related Work

Recently, there are several privacy-preserving data aggregation schemes proposed for smart grid communications, we briefly review some of them [3–11] in this section.

Based on the Paillier homomorphic encryption, Garcia and Jacobs [3] first introduce a privacy-friendly energy-metering scheme for smart grid communications, and then Erkin and Tsudik [4] propose a more flexible privacy-preserving scheme supporting both spatial and temporal power consumption aggregation. Chen et al. [5] combine Paillier encryption and secret sharing technique to propose a fault-tolerant privacy-preserving data aggregation. In addition, based on some homomorphic techniques, other efficient schemes [6–10] have also been proposed, some of them are even secure against differential attack [7–9]. However, all above schemes are the whole user set aggregation, and cannot support partially subset aggregation. In [11], Lu et al. propose an efficient and privacy-preserving aggregation scheme, called EPPA (i.e., PPMDA in Chap. 3), which uses a super-increasing sequence to structure multi-dimensional data and encrypt the structured data with Paillier encryption. Inherently, EPPA can support set aggregation in smart grid communications. However, if the CC is honest-but curious in EPPA, the CC can read each individual user's data.

Different from the above schemes, our proposed scheme supports subset data aggregation in smart grid communications, thus it can provide more accurate information for smart grid monitoring. In addition, even though the CC is honest-but curious, the CC still cannot read each individual user's data.

4.7 Summary

In this chapter, we have proposed a novel privacy-preserving subset data aggregation (PPSDA) scheme for smart grid communications. Given a threshold th of user electronic consumption data, the whole residential users $\mathbb{U}$ are divided into two subsets $\mathbb{U}_1$ and $\mathbb{U}_0$. The proposed scheme can just use one single aggregated ciphertext to aggregate the sum of electronic consumption data in each subset and the corresponding subset size, which thus supports more accurate data analytics for controlling and monitoring the smart grid. Detailed security analysis shows that the proposed scheme is privacy-preserving, i.e., no one can read each individual user's data, and only the CC can read the set aggregation results. Through extensive performance evaluations, we have also demonstrated that the proposed scheme is efficient in terms of computational costs and communication overhead.

In order to support more advanced data analytics, in the next chapter, we will introduce a privacy-preserving multi-functional data aggregation (MuDA) scheme for secure smart grid communications.

Appendix: A Sample Java Source Code of PPSDA

```
import java.math.BigInteger;
import java.security.MessageDigest;
import java.security.NoSuchAlgorithmException;
import java.security.SecureRandom;
import java.text.SimpleDateFormat;
import java.util.Date;
import java.util.Random;

/**
 * @ClassName: PPSDA
 * @Description: This is sample java code of PPSDA.
 */
public class PPSDA {
    private final int MAXUSER = 500; // max user
    private final int MAXRANGE = 10; // max range
    private final int CERTAINTY = 64;
    private int ModLength; // length in bits of the modulus n
    private BigInteger p; // a random prime
    private BigInteger q; // a random prime (distinct from p)
    private BigInteger n; // n=p*q
    private BigInteger pp; // a big prime pp=2*p*q+1
    private BigInteger gn; // generator of Group G with
            order n
    private BigInteger h0; // h0=g^q
    private BigInteger h1; // h1=g^p
    private BigInteger[] privatekeys; // private keys for users
    private BigInteger[] c; // ci=g^mi*h1*H(t)^xi
    private BigInteger d; // D
    private BigInteger d1; // D1
    private BigInteger d2; // D2
    private BigInteger ht; // hash(time)

    private int usersnumber; // user number
    private int[] usercom; // user consumption
    private int threshold; // threshold
    private int max; // the max electricity consumption of
            one user
    private int mi; // mi
    private int mj; // mj
    private int ui; // ui

    /**
     * @ClassName: Datai
     * @Description: This is a class for storing mi and ui.
     */
```

(continued)

```
public class Datai {
    private int mi;
    private int ui;

    public Datai(int mi, int ui) {
        this.mi = mi;
        this.ui = ui;
    }

    public int getMi() {
        return mi;
    }

    public int getUi() {
        return ui;
    }
}

/**
 * @ClassName: Dataj
 * @Description: This is a class for storing mj.
 */
public class Dataj {
    private int mj;

    public Dataj(int mj) {
        this.mj = mj;
    }

    public int getMj() {
        return mj;
    }
}

/**
 * @param ModLengthIn
 *           the security parameter, which decides the
 *           length of large prime pp.
 * @throws Exception
 *            If ModLengthIn<1024, there is an exception.
 */
public PPSDA(int ModLengthIn) throws Exception {
    if (ModLengthIn < 1024)
        throw new Exception("PPSDA(int ModLength): "
                + "Length must be >= 1024");

    mi = 0;
    mj = 0;
    ui = 0;
    ht = hashTime(getCurrentTimeStamp());
```

(continued)

```
        this.ModLength = ModLengthIn;
        generateKeys();
    }

    /**
     * @Title: generateKeys
     * @Description: This function is to generate keys.
     * @return void
     */
    public void generateKeys() {
        p = new BigInteger(ModLength / 2 + 2, CERTAINTY,
            new SecureRandom());
        do {
            q = new BigInteger(ModLength / 2, CERTAINTY,
                new SecureRandom());
            pp = BigInteger.valueOf(2).multiply(p).multiply(q)
                    .add(BigInteger.ONE);
            if (pp.isProbablePrime(CERTAINTY)) {
                BigInteger test = p.divide(q);
                if (test.compareTo(BigInteger.valueOf(2)) == 1)
                    break;
            }
        } while (true);
        n = p.multiply(q);
        gn = getGeneratorFromZn(pp, p, q);
        h0 = gn.modPow(q, pp);
        h1 = gn.modPow(p, pp);
        printallParameters();
    }

    /**
     * @Title: initializeParameters
     * @Description: This function is to initialize the
     *               related parameters.
     * @param usernumber
     *            The number of users.
     * @param threshold
     *            The threshold used to classify data.
     * @param max
     *            The maximum electricity consumption for
     *            one user.
     * @return void
     */
    public void initializeParameters(int usersnumber, int
        threshold, int max) {
        this.usersnumber = usersnumber;
        this.threshold = threshold;
        this.max = max;
        usercom = new int[this.usersnumber];
        privatekeys = new BigInteger[this.usersnumber + 1];
```

(continued)

```
        c = new BigInteger[this.usersnumber];
    }

    /**
     * @Title: getGeneratorFromZn
     * @Description: This function is to get a generator
     *               from Z_n.
     * @param pp
     *            A large prime, pp=2*p*q+1.
     * @param p
     *            A large prime.
     * @param q
     *            A large prime.
     * @return BigInteger A generator of Z_n.
     */
    private BigInteger getGeneratorFromZn(BigInteger pp,
            BigInteger p, BigInteger q) {
        BigInteger d;
        BigInteger a, b, c;
        BigInteger onetest = BigInteger.ONE;
        do {
            d = getRandomFromZpp();
            a = d.modPow(BigInteger.valueOf(2), pp);
            b = d.modPow(p, pp);
            c = d.modPow(q, pp);
            // satisfy the conditions
            if (!a.equals(onetest) && !b.equals(onetest)
                    && !c.equals(onetest)
                    && d.gcd(pp).equals(onetest)) {
                break;
            }
        } while (true);

        return d.modPow(BigInteger.valueOf(2), pp);
    }

    /**
     * @Title: getRandomFromZpp
     * @Description: This function is to get a random
     *               value from Z_pp.
     * @return BigInteger A random value in Z_pp.
     */
    private BigInteger getRandomFromZpp() {
        BigInteger r;
        do {
            r = new BigInteger(ModLength, new SecureRandom());
        } while (r.compareTo(BigInteger.ZERO) <= 0 ||
          r.compareTo(pp) >= 0);
        return r;
    }
```

(continued)

```
/**
 * @Title: getRandomFromZStarN
 * @Description: This function is to get a random value
 *               from Z*_n.
 * @return BigInteger A random value in Z*_n.
 */
private BigInteger getRandomFromZStarN() {
    BigInteger r;

    do {
        r = new BigInteger(ModLength, new SecureRandom());
    } while (r.compareTo(n) >= 0 || r.gcd(n)
      .intValue() != 1);
    return r;
}

/**
 * @Title: printallParameters
 * @Description: This function prints all related
 *               parameters
 * @return void
 */
private void printallParameters() {
    System.out.println("pp:" + pp.toString());
    System.out.println("p: " + p.toString());
    System.out.println("q: " + q.toString());
    System.out.println("n: " + n.toString());
    System.out.println("g: " + gn.toString());
    System.out.println("h0: " + h0.toString());
    System.out.println("h1: " + h1.toString());

}

/**
 * @Title: findDatai
 * @Description: This function use brute-force search to
 *               find the data.
 * @param c1
 *             The input data.
 * @return Datai The found data.
 * @throws Exception
 *              If the data is not found, there is
 *              an exception.
 */
private Datai findDatai(BigInteger c1) throws Exception {
    BigInteger exponent;
    Datai data = null;
    for (int i = 0; i <= MAXUSER; i++) {
        for (int j = 0; j <= MAXUSER * MAXRANGE; j++) {
            exponent = BigInteger.valueOf(j).add(
```

(continued)

```
                    p.multiply(BigInteger.valueOf(i)));
                if (c1.compareTo(h1.modPow(exponent,
                    pp)) == 0) {
                    data = new Datai(j, i);
                    return data;
                }
            }
        }
        throw new Exception("findDatai(BigInteger c1): "
                + "cannot find the data.");
    }

    /**
     * @Title: findDataj
     * @Description: This function use brute-force search
     *               to find the data.
     * @param c1
     *            The input data.
     * @return Dataj The found data.
     * @throws Exception
     *             If the data is not found, there is
     *             an exception.
     */
    private Dataj findDataj(BigInteger c2) throws Exception {
        for (int i = 0; i <= MAXUSER * MAXRANGE; i++) {
            if (c2.compareTo(h0.modPow(BigInteger.valueOf(i),
                pp)) == 0) {
                return new Dataj(i);
            }
        }
        throw new Exception("findDataj(BigInteger c2): "
                + "cannot find the data.");
    }

    /**
     * @Title: generatePrivateKeysForUsers
     * @Description: This function generates the private keys
     * for each user.
     * @return void
     */
    public void generatePrivateKeysForUsers() {

        BigInteger sum = BigInteger.ZERO;
        for (int i = 1; i < usersnumber + 1; i++) {
            privatekeys[i] = getRandomFromZStarN();
            sum = sum.add(privatekeys[i]);
        }
        privatekeys[0] = (BigInteger.ZERO.subtract(sum)).mod(n
                .multiply(BigInteger.valueOf(2)));
    }
```

(continued)

```
/**
 * @Title: getCurrentTimeStamp
 * @Description: This function returns the current
 *               time stamp.
 * @return String The current time stamp, "yyyy-MM-dd
 * HH:mm:ss"
 */
private String getCurrentTimeStamp() {
    SimpleDateFormat sdfDate = new SimpleDateFormat
    ("yyyy-MM-dd HH:mm:ss");
    Date now = new Date();
    String strDate = sdfDate.format(now);
    return strDate;
}

/**
 * @Title: hashTime
 * @Description: This function is a hash function,
 *               which is used to hash the time.
 * @param time
 *          The current time stamp.
 * @return BigInteger The hash value of time stamp.
 */
private BigInteger hashTime(String time) throws
    NoSuchAlgorithmException {
    MessageDigest md = MessageDigest
    .getInstance("SHA-256");
    md.update(time.getBytes());
    return new BigInteger(1, md.digest());
}

/**
 * @Title: generateUsersConsumption
 * @Description: This function is to generate the
 *               consumption of each user.
 * @return void
 */
public void generateUsersConsumption() {
    Random random = new Random();
    int realmi = 0;
    int realmj = 0;
    int realui = 0;
    for (int i = 0; i < usersnumber; i++) {
        usercom[i] = random.nextInt(max);
        if (usercom[i] <= threshold) {
            realui++;
            realmi = realmi + usercom[i];
        } else {
            realmj = realmj + usercom[i];
        }
```

(continued)

```
        }
        System.out.println("realmi, realmj, realui:
                " + realmi + " " + realmj
                + " " + realui);
    }

    /**
     * @Title: enc
     * @Description: This function simulates the process
     *               that user encrypt the
     *               data and upload them.
     * @return void
     */
    public void enc() {
        long enctime1 = System.currentTimeMillis();
        BigInteger tmp1; // g^mi
        BigInteger tmp2; // h1 h0^mi
        BigInteger tmp3; // H(t)^xi mod pp
        for (int i = 0; i < usersnumber; i++) {
            if (usercom[i] <= threshold) {
                tmp1 = gn.pow(usercom[i]);
                tmp2 = h1;
                tmp3 = ht.modPow(privatekeys[i + 1], pp);
                c[i] = (tmp1.multiply(tmp2).multiply(tmp3))
                .mod(pp);
            } else {
                tmp2 = h0.pow(usercom[i]);
                tmp3 = ht.modPow(privatekeys[i + 1], pp);
                c[i] = (tmp2.multiply(tmp3)).mod(pp);
            }
        }
        long enctime2 = System.currentTimeMillis();
        System.out.println("enc time cost: " + ((double)
                enctime2 - enctime1)/ usersnumber);
    }

    /**
     * @Title: dec
     * @Description: This function simulates the process that
     *               gateway receive
     *               these data and aggregate them, and then
     *               control center can
     *               decrypt them to get the final results.
     * @return void
     * @throws Exception
     *              If the encrypted data cannot be decrypted,
     *              there is an exception.
     */
    public void dec() throws Exception {
        long aggtime1 = System.currentTimeMillis();
```

(continued)

```
        BigInteger C = BigInteger.ONE;
        for (int i = 0; i < usersnumber; i++) {
            C = C.multiply(c[i]);
        }
        long aggtime2 = System.currentTimeMillis();
        System.out.println("agg cost time: "
                + (aggtime2 - aggtime1)
                + " usernumber: " + usersnumber);
        long dectime1 = System.currentTimeMillis();
        d = (C.mod(pp).multiply(ht.modPow(privatekeys[0],
        pp))).mod(pp);
        d1 = (d.modPow(p, pp)).mod(pp);
        Datai datai = findDatai(d1);
        mi = datai.getMi();
        ui = datai.getUi();
        d2 = d.multiply(gn.pow(mi).multiply(h1.pow(ui))
        .modInverse(pp)).mod(pp);
        Dataj dataj = findDataj(d2);
        mj = dataj.getMj();
        System.out.println("mi, mj, ui: " + mi + " " + mj
        + " " + ui);
        long dectime2 = System.currentTimeMillis();
        System.out.println("dec cost time: " + (dectime2
               - dectime1)+ " usernumber: " + usersnumber);
    }

    // Simulation.
    public static void main(String[] args) {
        try {
            PPSDA test = new PPSDA(1024);
            int usernumber = 10;
            int maxelec = 10;
            System.out.println(">>>>>>>>>>>>>>>>>>>>>>>>>>
                  >>>>>>>>>>>>>");
            int threshold = new Random().nextInt(10);
            System.out.println("threshold: " + threshold);
            test.initializeParameters(usernumber, threshold,
                maxelec);
            test.generatePrivateKeysForUsers();
            test.generateUsersConsumption();
            test.enc();
            test.dec();
        } catch (Exception e) {
            e.printStackTrace();
        }
    }
}
```

References

1. R. Lu, K. Alharbi, X. Lin, and C. Huang, "A novel privacy-preserving set aggregation scheme for smart grid communications," in *Proc. of IEEE Globecom 2015*, San Diego, CA, USA, December 2015.
2. "Consumer benefits," http://www.whatissmartgrid.org/smart-grid-101/consumer-benefits.
3. F. D. Garcia and B. Jacobs, "Privacy-friendly energy-metering via homomorphic encryption," in *Security and Trust Management - 6th International Workshop, STM 2010, Athens, Greece, September 23–24, 2010, Revised Selected Papers*, 2010, pp. 226–238.
4. Z. Erkin and G. Tsudik, "Private computation of spatial and temporal power consumption with smart meters," in *Applied Cryptography and Network Security - 10th International Conference, ACNS 2012, Singapore, June 26–29, 2012. Proceedings*, 2012, pp. 561–577.
5. L. Chen, R. Lu, and Z. Cao, "PDAFT: A privacy-preserving data aggregation scheme with fault tolerance for smart grid communications," *Peer-to-Peer Networking and Applications (PPNA) (Springer)*, to appear.
6. K. Alharbi and X. Lin, "LPDA: A lightweight privacy-preserving data aggregation scheme for smart grid," in *International Conference on Wireless Communications and Signal Processing, WCSP 2012, Huangshan, China, October 25–27, 2012*, 2012, pp. 1–6. [Online]. Available: http://dx.doi.org/10.1109/WCSP.2012.6542936
7. E. Shi, T. H. Chan, E. G. Rieffel, R. Chow, and D. Song, "Privacy-preserving aggregation of time-series data," in *Proceedings of the Network and Distributed System Security Symposium, NDSS 2011, San Diego, California, USA, 6th February - 9th February 2011*, 2011.
8. H. Bao and R. Lu, "A new differentially private data aggregation with fault tolerance for smart grid communications," *IEEE Internet of Things Journal*, to appear.
9. L. Chen, R. Lu, Z. Cao, K. AlHarbi, and X. Lin, "MuDA: Multifunctional data aggregation in privacy-preserving smart grid communications," *Peer-to-Peer Networking and Applications (PPNA) (Springer)*, to appear.
10. C. Li, R. Lu, H. Li, L. Chen, and J. Chen, "PDA: A privacy-preserving dual-functional aggregation scheme for smart grid communications," *Security and Communication Networks*, to appear.
11. R. Lu, X. Liang, X. Li, X. Lin, and X. Shen, "EPPA: an efficient and privacy-preserving aggregation scheme for secure smart grid communications," *IEEE Trans. Parallel Distrib. Syst.*, vol. 23, no. 9, pp. 1621–1631, 2012. [Online]. Available: http://dx.doi.org/10.1109/TPDS.2012.86
12. Y. Liu, M. K. Reiter, and P. Ning, "False data injection attacks against state estimation in electric power grids," in *Proceedings of the 2009 ACM Conference on Computer and Communications Security, CCS 2009, Chicago, Illinois, USA, November 9–13, 2009*, 2009, pp. 21–32. [Online]. Available: http://doi.acm.org/10.1145/1653662.1653666
13. D. Boneh, E. Goh, and K. Nissim, "Evaluating 2-dnf formulas on ciphertexts," in *Theory of Cryptography, Second Theory of Cryptography Conference, TCC 2005, Cambridge, MA, USA, February 10–12, 2005, Proceedings*, 2005, pp. 325–341.
14. A. J. Menezes, P. C. Van Oorschot, and S. A. Vanstone, *Handbook of applied cryptography*. CRC press, 1997.
15. M. Bellare and P. Rogaway, "Random oracles are practical: A paradigm for designing efficient protocols," in *CCS '93, Proceedings of the 1st ACM Conference on Computer and Communications Security, Fairfax, Virginia, USA, November 3–5, 1993.*, 1993, pp. 62–73. [Online]. Available: http://doi.acm.org/10.1145/168588.168596
16. R. Lu, X. Lin, Z. Cao, J. Shao, and X. Liang, "New (t, n) threshold directed signature scheme with provable security," *Inf. Sci.*, vol. 178, no. 3, pp. 756–765, 2008. [Online]. Available: http://dx.doi.org/10.1016/j.ins.2007.07.025

Chapter 5
Privacy-Preserving Multifunctional Data Aggregation

In this chapter, we discuss a new multifunctional data aggregation scheme, named MuDA, for privacy-preserving smart grid communications [1]. With MuDA, the smart grid control center can compute multiple statistic functions of users' data in a privacy-preserving way to provide diversiform services.

5.1 Introduction

The 2006 European blackout was a major blackout originated in Germany, but cascaded across Europe extending from Poland to the Benelux countries, France, Portugal, Spain, etc. One primary cause of this blackout is due to the insufficient inner-communication of the Transmission System Operators, and most of the operators did not have access to real-time data [2]. Moreover, investigations of the 2003 North America blackout also show that the malfunction was due to load imbalance in the electric power grid and lack of effective real-time diagnosis [3]. These incidents obviously demonstrate that the traditional power grid is undoubtedly outdated and no longer meets our growing demand for continuous real-time monitoring and stable electricity distribution. Due to this reason, in the recent decade, ever-increasing efforts on the development of next-generation power grid, known as *smart grid*, have been made in many countries all over the world [4, 5].

Compared with traditional power grid, smart grid has introduced new concepts and promising solutions for intelligent electricity generation, transmission, distribution and utilization. By deploying various sensors along with the two-way flows of electricity and communication, a huge amount of real-time data are collected and reported to the *control center* (CC) for timely monitoring and additional analysis, as illustrated in Fig 5.1. With the received information as feedback, the CC can automatically and timely monitor grid status, balance electricity load, maintain system operation, optimize energy consumption, etc. [6]. In some applications,

R. Lu, *Privacy-Enhancing Aggregation Techniques for Smart Grid Communications*, Wireless Networks, DOI 10.1007/978-3-319-32899-7_5

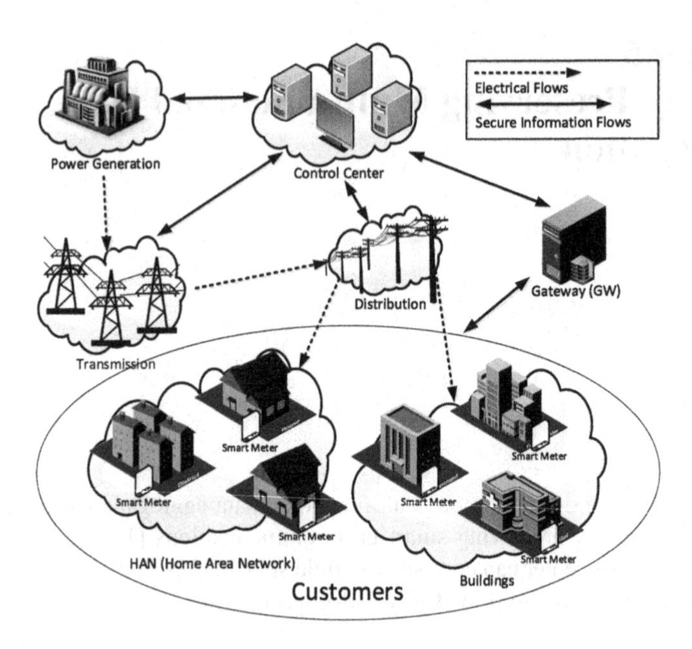

Fig. 5.1 Conceptual smart grid system architecture

the CC needs to compute certain statistics of residential users' electricity usage, e.g., summation, average, variance and so on. Specifically, all the intelligent electric appliances in the residential user's home are connected to a key element, *smart meter*, which periodically records the power consumption of appliances and reports the metering data to a local area *gateway* (GW), e.g., a workstation with wireless access modules. The GW then collects, preprocesses (e.g., authenticates, aggregates) and forwards the data to the CC for further analyzing and processing.

However, the real-time user data, e.g., collected every 15 min, contain specific power usage patterns which are highly relevant to users' private lives, thus they must be protected from unauthorized entities. Existing data aggregation schemes [7–12] also stress the same consideration that during aggregating, individual user's data should not (actually need not) be exposed. Most of them use a homomorphic encryption to encrypt users' data so that the semi-trust aggregator (e.g. the GW) can aggregate all users' data without decryption. However, all of these schemes can only be used to compute the summation of users' data as the aggregation, while the CC may need to compute more statistics such as the mostly common used *variance*. Therefore, how to compute more complex statistics of users' data without disclosing individual user's data is a challenging problem in smart grid communications. Actually, this problem is also mentioned as one of the open research challenges in [11].

Another challenging problem that each secure data aggregation scheme could face is the differential attack [13]. The idea of differential attack is straight forward. Even if an aggregation scheme is secure, once the CC acquires the summation of n

users and that of $n-1$ users, the privacy of the "differential" one is leaked, although the aggregation of n users and that of $n-1$ users are both secure. This problem has been addressed in several literatures, such as [7, 11, 13–15]. However, all of them just consider protecting differential privacy upon summation aggregation. To the best of our knowledge, we are the first to consider the differential privacy of variance aggregation and other more complex aggregations such as *one-way ANOVA* (ANalysis Of VAriance) [16] aggregation.

In order to assist the CC to compute more complex statistics in a privacy-preserving way, in this chapter we propose a novel *Mu*ltifunctional *D*ata *A*ggregation scheme, named MuDA, in the smart grid communications. MuDA achieves privacy-preserving aggregation of multiple functions such as average, variance, one-way ANOVA, etc. In our enhanced version of MuDA, we additionally protect users' privacy against differential attacks upon multifunctional aggregations. Specifically, the main contributions of this chapter are three-fold.

- Firstly, since the CC may need to compute multiple statistic functions to provide diversiform services, we present a novel MuDA scheme that supports multifunctional aggregations in smart grid communications. Compared with earlier data aggregation schemes that can only compute summation aggregation, our proposed scheme provides more diversity and flexibility for the CC.
- Secondly, inspired by the fact that users' private data may also suffer from the differential attacks, our enhanced version of MuDA is designed to provide differential privacy upon multifunctional aggregations with limited noises.
- Thirdly, we optimize our basic scheme to reduce the computation complexity of the CC while introducing a little more communication cost. In addition, through comparative performance analysis, we demonstrate that MuDA is more efficient than a popular aggregation scheme [11] in terms of communication overhead.

The remainder of this chapter is organized as follows. In Sect. 5.2, we introduce the system model, security model and design goal. Then, our basic MuDA scheme and the enhanced version are introduced in Sects. 5.3 and 5.4 respectively, followed by the security analysis and utility analysis in Sect. 5.5. We also compare the performance of our scheme with a popular aggregation scheme [11] in Sect. 5.6 and discuss other related work in Sect. 5.7. Finally, we draw our conclusions in Sect. 5.8.

5.2 Problem Formalization

In this section, we formalize our research problems on multifunctional data aggregations in smart grid communications, including system model, security model and design goal.

5.2.1 System Model

Since residential users always care about their privacy when reporting their detailed electricity usage data to the control center in smart grid communication, in this work, we mainly focus on how to let the control center compute multiple statistic functions of users' data, such as average, variance, one-way ANOVA, etc., in a privacy-preserving way. Specifically, in our system model, we consider a typical smart grid communication architecture for residential users, which consists of a trusted authority (TA), a control center (CC) in charge of communication and control of the system, a local gateway (GW), and a large number of residential users $\mathbb{U} = \{U_1, U_2, \cdots, U_n\}$ in a residential area (RA), as shown in Fig. 5.2. The local GW is a powerful workstation which connects the CC and residential users, i.e., assisting the CC in collecting residential users' nearly real-time electricity usage data. The communication between residential users and the GW is via WiFi technology as suggested in the Standards [17], while the communication between the GW and CC is via wired links with high bandwidth and low delay. The responsibility of GW in our system is mainly twofold, one is collecting and relaying users' data, and the other is performing some aggregations based on the CC's requirements.

Multiple Statistic Functions of Data Aggregation. For the purpose of providing diversified services, the CC needs to compute multiple statistics or aggregations of users' data. For example, by computing the summation of users' data, the CC can make real-time power pricing [18], detect power fraud/leakage [19], forecast electricity usage, etc.; by computing the variance of users' data, the CC can understand the uniformity of electricity usage distribution, helping to detect load imbalance and abnormal situations; by computing more complex statistical models such as one-way ANOVA (analysis of variance [16], which is a particular form of statistical hypothesis testing commonly used in the analysis of experimental data), the CC could make decisions precisely based on the statistical results, e.g. whether

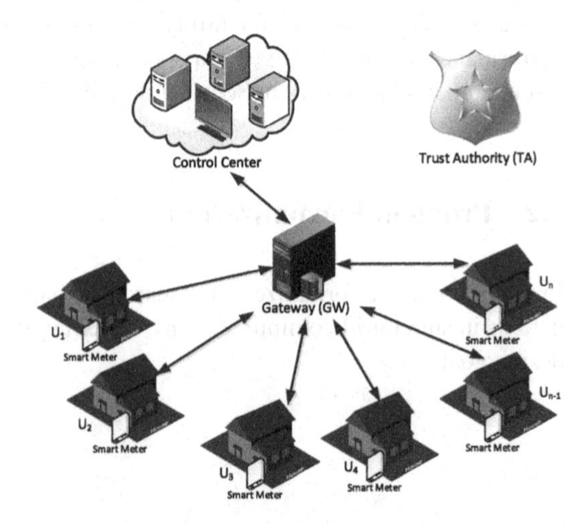

Fig. 5.2 System model under consideration

its new power pricing strategies have statistically significant influence on users' electricity usage. By allowing the GW to perform multifunctional aggregations, our system provides rich diversity of statistic data for the CC to improve its service qualities in the smart grid.

5.2.2 Security Model

While reporting fine-grained electricity usage data, users are also worried about leaking privacy of their activities. In our security model, we consider preventing the adversary from revealing the individual user's electricity usage data, and at the same time allowing the CC to compute multiple functions upon users' data. Specifically, we consider the CC and GW are both trustable, and the users $\mathbb{U} = \{U_1, U_2, \cdots, U_n\}$ are all honest. However, their exists a malicious adversary $\mathscr{A}$ eavesdropping the communication flows between users and the GW, and those between the GW and CC. In addition, the adversary $\mathscr{A}$ could intrude into the databases of GW and CC to steal the stored data. More seriously, based on the aggregated data obtained, the adversary $\mathscr{A}$ may launch some differential attacks to acquire the individual user's data. Therefore, we consider achieving the following security requirements to avoid the adversary $\mathscr{A}$ from disclosing individual user's sensitive data in smart grid communications. Note that we mainly focus on protecting user data privacy in this chapter, ensuring integrity of user data is out of the scope of our work. In fact, by adding some authentication techniques at the GW, the integrity of users' data could be ensured.

- *Data Confidentiality.* As the adversary $\mathscr{A}$ may reside in the RA to eavesdrop the communication flows, data transmitted from residential users to the GW and that from the GW to CC should both be encrypted. In order to aggregate users' data, the GW could first decrypt the data reported from users, aggregate them together, and then encrypt the aggregation again and transmit it to the CC. However, this way may cost much time on decrypting all users' data. More seriously, if the adversary $\mathscr{A}$ intrudes into the database of GW, then all users' private data is disclosed. As a result, the GW should be able to aggregate users' data in a privacy-preserving way, i.e. without decryption. Then the confidentiality of users' private data can be ensured.
- *Differential Privacy.* Although some powerful adversary may also intrude into the database of CC, the individual user's data will not be disclosed directly since the CC only stores aggregated data. However, as we allow the CC to compute multifunctional aggregations, the adversary $\mathscr{A}$ may utilize multiple aggregations of similar sets to launch the differential attack and obtain the individual user's private data. Therefore, providing differential privacy of users' data is also required in the secure smart grid communications.

5.2.3 Design Goal

Under the aforementioned system model and security requirements, our design goal is to develop an efficient multifunctional data aggregation scheme while preserving user data privacy in smart grid communications. Specifically, the following three objectives should be achieved.

- *The computation of multifunctional aggregation should be allowed in the proposed aggregation scheme.* In order to provide diversified service for users, the CC may need to compute multiple functions of users' data, e.g. average, variance, one-way ANOVA, etc. To preserve user privacy, those functions should be computed in the form of ciphertext. Therefore, the proposed scheme should allow the GW to compute multifunctional aggregation without decryption.
- *The security requirements should be met in the proposed aggregation scheme.* As aforementioned, without protecting data privacy, users will worry about reporting their detailed electricity data to the CC, and the smart grid systems cannot step into its flourish. Therefore, the above security requirements should be satisfied in the proposed aggregation scheme.
- *The efficiency of communication should be achieved in the proposed aggregation scheme.* To compute multiple functions of user data, we can simply ask users to pre-compute some functions of their data, e.g. squares, and transmit the results to the GW. However, this way can only support limited and fixed functions, meanwhile bringing additional computation and communication cost to the users. Therefore, the proposed scheme should provide diversity of aggregation functions in an efficient way.

5.3 The Basic MuDA Scheme

In this section, we propose our basic multifunctional data aggregation scheme, called MuDA, for smart grid communications. It mainly consists of the following four parts: system initialization, user report generation, privacy-preserving report aggregation and secure report reading. The basic scheme mainly focuses on providing multifunctional aggregation, i.e. average, variance, and one-way ANOVA. We show that our scheme can efficiently compute those statistical aggregations in a privacy-preserving way, but not limited on them. Our enhanced version can additionally support differential privacy to provide advanced security level, which will be expanded in the next section.

5.3.1 *System Initialization*

At the beginning, the single trusted authority (TA) can bootstrap the whole system. Specifically, in the system initialization phase, given the security parameters τ, TA first runs the algorithm $\mathcal{CG}en(\tau)$ to generate the bilinear map tuple $(p, q, \mathbb{G}, \mathbb{G}_T, e)$. Then TA builds up the BGN PKE, acquires the tuple $(N, \mathbb{G}, \mathbb{G}_T, e, g, h)$, where $N = pq$, $g \in \mathbb{G}$ is a random generator of $\mathbb{G}$, and $h = g^{q\beta}$ (for some private β) is a random generator of the subgroup of $\mathbb{G}$ of order p. TA also chooses a secure cryptographic hash function $H : \{0, 1\}^* \rightarrow \mathbb{G}$. Finally, TA publishes $(N, \mathbb{G}, \mathbb{G}_T, e, g, h, H)$ as the public key of our system, and assigns the private key $SK = p$ to the CC through a secure channel. Although some private keys may also be assigned to users for the sake of authentication, since it is not our main focus, we do not consider it in this work.

5.3.2 *User Report Generation*

Assume the reporting time points of our system, e.g. every 15 min, are defined as $\mathbb{T} = \{t_1, t_2, \ldots, t_{max}\}$ for a sufficient long runtime period. By reporting nearly real-time electricity usage data at these time points simultaneously, users can help the CC monitor the health of smart grid and provide various services. Specifically, each user $U_i \in \mathbb{U}$ collects its usage data $m_{i,\gamma} \in \{0, 1, \ldots, W\}$ at time point t_γ, and performs the following steps:

Step-1: U_i first computes the hash value $g_\gamma = H(t_\gamma)$ for the current reporting time point t_γ.

Step-2: Then, U_i chooses a random number $r_{i,\gamma} \in \mathbb{Z}_N$, and computes

$$C_{i,\gamma} = g_\gamma^{m_{i,\gamma}} \cdot h^{r_{i,\gamma}}. \tag{5.1}$$

Step-3: Eventually, U_i reports $C_{i,\gamma}$ to the GW through WiFi.

Note that for residential users, the electricity usage within 15 min could not be extremely high, thus choosing an appropriate large W is enough in reality.

5.3.3 *Privacy-Preserving Report Aggregation*

After receiving all n encrypted usage data $C_{i,\gamma}$ for $i = 1, 2, \ldots, n$, the GW could aggregate users' data based on the CC's requirements in a privacy-preserving way. Define the aggregation of users' data as a function $\mathrm{A} : \{0, 1, \ldots, W\}^n \rightarrow \mathbb{Z}_q$ which takes users' data $(m_1, m_2, \ldots, m_n)$ as input and the aggregated result as output. Then our basic scheme can support multiple aggregation functions depending on the CC's requirements. We illustrate three kinds of aggregations which are used frequently in statistics, i.e. average, variance and one-way ANOVA. Note that our scheme is not limited in the aforementioned aggregations. In fact, for any given aggregation function that has the maximum degree of 2, the GW could aggregate the users' data in a privacy-preserving way by using our scheme.

5.3.3.1 Average Aggregation

If the CC wants to compute the average usage of users, it sends the aggregation function $\mathrm{A}_{ave} = \frac{1}{n}\sum_{i=1}^{n} m_i$ to the GW. When the users' electricity usage data are received, the GW can compute the aggregation with the following steps.

Step-1: The GW first computes the encrypted aggregation $\mathrm{A}_{1,\gamma}$ from users' data $C_{1,\gamma}, C_{2,\gamma}, \ldots, C_{n,\gamma}$ as

$$\begin{aligned} \mathrm{A}_{1,\gamma} &= \prod_{i=1}^{n} C_{i,\gamma} = \prod_{i=1}^{n} (g_\gamma^{m_{i,\gamma}} \cdot h^{r_{i,\gamma}}) \\ &= g_\gamma^{\sum_{i=1}^{n} m_{i,\gamma}} \cdot h^{R_1}, \end{aligned} \tag{5.2}$$

where $R_1 = \sum_{i=1}^{n} r_{i,\gamma} \mod N$.

Step-2: Then the GW reports the aggregated data $\mathrm{A}_{1,\gamma}$ to the CC for further computation.

5.3.3.2 Variance Aggregation

If the CC wants to compute the variance of users' electricity usage, it sends the aggregation function $\mathrm{A}_{var} = \sum_{i=1}^{n}(m_i - \bar{m})^2$ to the GW, where $\bar{m}$ denotes the average of m_i. When the users' electricity usage data are received, the GW can compute the aggregations with the following steps.

Step-1: The GW first computes the encrypted aggregation $\mathrm{A}_{2,\gamma}$ from users' data $C_{1,\gamma}, C_{2,\gamma}, \ldots, C_{n,\gamma}$ as

$$\begin{aligned} \mathrm{A}_{2,\gamma} &= e(\prod_{i=1}^{n} C_{i,\gamma}, \prod_{i=1}^{n} C_{i,\gamma}) \\ &= e(\prod_{i=1}^{n} (g_\gamma^{m_{i,\gamma}} \cdot h^{r_{i,\gamma}}), \prod_{i=1}^{n} (g_\gamma^{m_{i,\gamma}} \cdot h^{r_{i,\gamma}})) \\ &= e(g_\gamma^{\sum_{i=1}^{n} m_{i,\gamma}} \cdot h^{\sum_{i=1}^{n} r_{i,\gamma}}, g_\gamma^{\sum_{i=1}^{n} m_{i,\gamma}} \cdot h^{\sum_{i=1}^{n} r_{i,\gamma}}) \\ &= e(g_\gamma, g_\gamma)^{(\sum_{i=1}^{n} m_{i,\gamma})^2} \cdot e(g_\gamma, h)^{R_2}, \end{aligned} \tag{5.3}$$

where $R_2 = 2\sum_{i=1}^{n} m_{i,\gamma} \cdot \sum_{i=1}^{n} r_{i,\gamma} + \beta q(\sum_{i=1}^{n} r_{i,\gamma})^2 \mod N$.

Step-2: And then the GW computes

$$\begin{aligned} A_{3,\gamma} &= \prod_{i=1}^{n} e(C_{i,\gamma}, C_{i,\gamma}) \\ &= \prod_{i=1}^{n} e(g_\gamma, g_\gamma)^{m_{i,\gamma}^2} \cdot \prod_{i=1}^{n} e(g_\gamma, h)^{2m_{i,\gamma} r_{i,\gamma} + \beta q r_{i,\gamma}^2} \\ &= e(g_\gamma, g_\gamma)^{\sum_{i=1}^{n} m_{i,\gamma}^2} \cdot e(g_\gamma, h)^{R_3}, \end{aligned} \tag{5.4}$$

where $R_3 = \sum_{i=1}^{n}(2m_{i,\gamma} r_{i,\gamma} + \beta q r_{i,\gamma}^2) \mod N$.

Step-3: After that, the GW reports the aggregated data $(A_{2,\gamma}, A_{3,\gamma})$ to the CC for further computation.

5.3.3.3 One-Way ANOVA Aggregation

If the CC wants to compute the one-way ANOVA of users' electricity usage (e.g. the power supplier has $s \geq 3$ pricing strategies, in order to check whether these strategies have significant influence on users' electricity usage, the CC could test the one-way ANOVA of users' daily electricity usage data to find out the answer), it sends the aggregation functions $A_{SS_B} = \sum_{j=1}^{s}\sum_{i=1}^{n} m_{ij}^2 - \frac{1}{n}\sum_{j=1}^{s}(\sum_{i=1}^{n} m_{ij})^2$ and $A_{SS_W} = \frac{1}{n}\sum_{j=1}^{s}(\sum_{i=1}^{n} m_{ij})^2 - \frac{1}{ns}(\sum_{j=1}^{s}\sum_{i=1}^{n} m_{ij})^2$ to the GW, where m_{ij} denotes the daily electricity usage of U_i under the jth pricing strategy and let the encryption of m_{ij} be $C_{ij} = g^{m_{ij}} \cdot h^{r_{ij}}$. When the users' encrypted electricity usage data C_{ij} are all received, the GW can compute the aggregations with the following steps.

Step-1: The GW first computes the encrypted aggregation A_4 from users' data C_{ij} as

$$\begin{aligned} A_4 &= \prod_{j=1}^{s}\prod_{i=1}^{n} e(C_{ij}, C_{ij}) \\ &= \prod_{j=1}^{s}\prod_{i=1}^{n} e(g, g)^{m_{ij}^2} \cdot \prod_{j=1}^{s}\prod_{i=1}^{n} e(g, h)^{2m_{ij} r_{ij} + \beta q r_{ij}^2} \\ &= e(g, g)^{\sum_{j=1}^{s}\sum_{i=1}^{n} m_{ij}^2} \cdot e(g, h)^{R_4}, \end{aligned} \tag{5.5}$$

where $R_4 = \sum_{j=1}^{s}\sum_{i=1}^{n}(2m_{ij} r_{ij} + \beta q r_{ij}^2) \mod N$.

Step-2: And then the GW computes

$$
\begin{aligned}
A_5 &= \prod_{j=1}^{s} e(\prod_{i=1}^{n} C_{ij}, \prod_{i=1}^{n} C_{ij}) \\
&= \prod_{j=1}^{s} e(g^{\sum_{i=1}^{n} m_{ij}} \cdot h^{\sum_{i=1}^{n} r_{ij}}, g^{\sum_{i=1}^{n} m_{ij}} \cdot h^{\sum_{i=1}^{n} r_{ij}}) \\
&= \prod_{j=1}^{s} e(g,g)^{(\sum_{i=1}^{n} m_{ij})^2}. \\
& \prod_{j=1}^{s} e(g,h)^{2\sum_{i=1}^{n} m_{ij} \cdot \sum_{i=1}^{n} r_{ij} + \beta q(\sum_{i=1}^{n} r_{ij})^2} \\
&= e(g,g)^{\sum_{j=1}^{s}(\sum_{i=1}^{n} m_{ij})^2} \cdot e(g,h)^{R_5},
\end{aligned}
\tag{5.6}
$$

where $R_5 = \sum_{j=1}^{s}(2\sum_{i=1}^{n} m_{ij} \cdot \sum_{i=1}^{n} r_{ij} + \beta q(\sum_{i=1}^{n} r_{ij})^2) \mod N$.

Step-3: After that, the GW calculates

$$
\begin{aligned}
A_6 &= e(\prod_{j=1}^{s}\prod_{i=1}^{n} C_{ij}, \prod_{j=1}^{s}\prod_{i=1}^{n} C_{ij}) \\
&= e(\bar{C}, \bar{C}) \qquad (\bar{C} = g^{\sum_{j=1}^{s}\sum_{i=1}^{n} m_{ij}} \cdot h^{\sum_{j=1}^{s}\sum_{i=1}^{n} r_{ij}}) \\
&= e(g,g)^{(\sum_{j=1}^{s}\sum_{i=1}^{n} m_{ij})^2} \cdot e(g,h)^{R_6},
\end{aligned}
\tag{5.7}
$$

where $R_6 = 2\sum_{j=1}^{s}\sum_{i=1}^{n} m_{ij} \cdot \sum_{j=1}^{s}\sum_{i=1}^{n} r_{ij} + \beta q(\sum_{j=1}^{s}\sum_{i=1}^{n} r_{ij})^2 \mod N$.

Step-4: Eventually, the GW reports the aggregated data (A_4, A_5, A_6) to the CC for further computation.

5.3.4 Secure Report Reading

Upon receiving the corresponding aggregations, the CC can efficiently compute the statistics needed while preserving users' privacy. Specifically, according to the aggregation CC requires, the CC makes the following calculations.

5.3.4.1 Average Aggregation

After receiving the aggregation $A_{1,\gamma} = g_\gamma^{\sum_{i=1}^n m_{i,\gamma}} \cdot h^{R_1}$ from the GW, the CC first computes $(A_{1,\gamma})^p$ by using the private key $SK = p$.

$$\begin{aligned} A_{1,\gamma}^p &= (g_\gamma^{\sum_{i=1}^n m_{i,\gamma}})^p \cdot (h^{R_1})^p \\ &= (g_\gamma^p)^{\sum_{i=1}^n m_{i,\gamma}} = \hat{g}_\gamma^{\sum_{i=1}^n m_{i,\gamma}}, \end{aligned} \tag{5.8}$$

where $\hat{g}_\gamma = g_\gamma^p$.

Since $m_{i,\gamma} \in \{1, 2, \ldots, W\}$, we have $\sum_{i=1}^n m_{i,\gamma} \leq nW$. By computing the discrete log of $\hat{g}_\gamma^{\sum_{i=1}^n m_{i,\gamma}}$ base $\hat{g}_\gamma$, the CC can get the sum of users' data $M_{sum} = \sum_{i=1}^n m_{i,\gamma}$ in expected time $O(\sqrt{nW})$ using Pollard's lambda method [20]. Then computing the average of users' data is straight forward.

$$M_{ave} = \frac{1}{n} M_{sum},$$

while the individual usage data of user are hidden.

5.3.4.2 Variance Aggregation

While receiving the aggregations $(A_{2,\gamma}, A_{3,\gamma})$ from the GW, the CC first computes $A_{2,\gamma}^p$ and $A_{3,\gamma}^p$ by using the private key $SK = p$. Specifically,

$$\begin{aligned} A_{2,\gamma}^p &= (e(g_\gamma, g_\gamma)^{(\sum_{i=1}^n m_{i,\gamma})^2})^p \cdot (e(g_\gamma, h)^{R_2})^p \\ &= (e(g_\gamma, g_\gamma)^p)^{(\sum_{i=1}^n m_{i,\gamma})^2} = \bar{g}_\gamma^{(\sum_{i=1}^n m_{i,\gamma})^2}, \end{aligned} \tag{5.9}$$

where $\bar{g}_\gamma = e(g_\gamma, g_\gamma)^p$. By computing the discrete log of $\bar{g}_\gamma^{(\sum_{i=1}^n m_{i,\gamma})^2}$ base $\bar{g}_\gamma$, the CC can acquire the square of sum of users' data $M_{sqrsum} = (\sum_{i=1}^n m_{i,\gamma})^2$ in expected time $O(nW)$.

By computing

$$\begin{aligned} A_{3,\gamma}^p &= (e(g_\gamma, g_\gamma)^{\sum_{i=1}^n m_{i,\gamma}^2})^p \cdot (e(g_\gamma, h)^{R_3})^p \\ &= (e(g_\gamma, g_\gamma)^p)^{\sum_{i=1}^n m_{i,\gamma}^2} = \bar{g}_\gamma^{\sum_{i=1}^n m_{i,\gamma}^2}, \end{aligned} \tag{5.10}$$

and the discrete log of $\bar{g}_\gamma^{\sum_{i=1}^n m_{i,\gamma}^2}$ base $\bar{g}_\gamma$, the CC can get the sum of squares of users' data $M_{sumsqr} = \sum_{i=1}^n m_{i,\gamma}^2$ in expected time $O(\sqrt{n}W)$. Then the variance M_{var} of users' data can be computed as

$$M_{var} = \frac{1}{n} M_{sumsqr} - \frac{1}{n^2} M_{sqrsum},$$

while the individual usage data of user are hidden.

5.3.4.3 One-Way ANOVA Aggregation

While receiving the aggregations (A_4, A_5, A_6) from the GW, the CC first computes A_4^p, A_5^p and A_6^p respectively by using the private key $SK = p$.

$$\begin{aligned} A_4^p &= (e(g,g)^{\sum_{j=1}^{s}\sum_{i=1}^{n} m_{ij}^2})^p \cdot (e(g,h)^{R_4})^p \\ &= (e(g,g)^p)^{\sum_{j=1}^{s}\sum_{i=1}^{n} m_{ij}^2} = \bar{g}^{\sum_{j=1}^{s}\sum_{i=1}^{n} m_{ij}^2}, \end{aligned} \tag{5.11}$$

where $\bar{g} = e(g,g)^p$. By computing the discrete log of $\bar{g}^{\sum_{j=1}^{s}\sum_{i=1}^{n} m_{ij}^2}$ base $\bar{g}$, the CC could first get the sum of squares of all data $M_4 = \sum_{j=1}^{s}\sum_{i=1}^{n} m_{ij}^2$ in expected time $O(\sqrt{ns}W)$.

$$\begin{aligned} A_5^p &= (e(g,g)^{\sum_{j=1}^{s}(\sum_{i=1}^{n} m_{ij})^2})^p \cdot (e(g,h)^{R_5})^p \\ &= (e(g,g)^p)^{\sum_{j=1}^{s}(\sum_{i=1}^{n} m_{ij})^2} = \bar{g}^{\sum_{j=1}^{s}(\sum_{i=1}^{n} m_{ij})^2}, \end{aligned} \tag{5.12}$$

By computing the discrete log of $\bar{g}^{\sum_{j=1}^{s}(\sum_{i=1}^{n} m_{ij})^2}$ base $\bar{g}$, the CC could then acquire the sum of squares of sums of data within each pricing strategy $M_5 = \sum_{j=1}^{s}(\sum_{i=1}^{n} m_{ij})^2$ in expected time $O(\sqrt{s}nW)$.

$$\begin{aligned} A_6^p &= (e(g,g)^{(\sum_{j=1}^{s}\sum_{i=1}^{n} m_{ij})^2})^p \cdot (e(g,h)^{R_6})^p \\ &= (e(g,g)^p)^{(\sum_{j=1}^{s}\sum_{i=1}^{n} m_{ij})^2} = \bar{g}^{(\sum_{j=1}^{s}\sum_{i=1}^{n} m_{ij})^2}, \end{aligned} \tag{5.13}$$

By computing the discrete log of $\bar{g}^{(\sum_{j=1}^{s}\sum_{i=1}^{n} m_{ij})^2}$ base $\bar{g}$, the CC could also get the square of sum of all data $M_6 = (\sum_{j=1}^{s}\sum_{i=1}^{n} m_{ij})^2$ in expected time $O(nsW)$.

According to the one-way ANOVA, the sum of squares between groups (pricing strategies) is $A_{SS_B} = \sum_{j=1}^{s}\sum_{i=1}^{n} m_{ij}^2 - \frac{1}{n}\sum_{j=1}^{s}(\sum_{i=1}^{n} m_{ij})^2$, and the sum of squares within groups is $A_{SS_W} = \frac{1}{n}\sum_{j=1}^{s}(\sum_{i=1}^{n} m_{ij})^2 - \frac{1}{ns}(\sum_{j=1}^{s}\sum_{i=1}^{n} m_{ij})^2$, thus

$$\begin{aligned} A_{SS_B} &= M_4 - \frac{1}{n} M_5 \\ A_{SS_W} &= \frac{1}{n} M_5 - \frac{1}{ns} M_6, \end{aligned} \tag{5.14}$$

After computing A_{SS_B} and A_{SS_W}, the CC could finally compute the F value of F-test as $F = \frac{A_{SS_W}/(s-1)}{A_{SS_B}/(n-s)}$. If the F value is greater than the F critical value with $s-1$ numerator and $n-s$ denominator degrees of freedom, the CC rejects the null hypothesis, which means at least one of the pricing strategies has significant influence on the electricity usage of users; otherwise the CC accepts the null hypothesis, which means no pricing strategy makes significant influence on users' electricity usage.

5.4 The Enhanced MuDA Version

In this section we propose the enhanced version of MuDA scheme which additionally protects user's electricity usage privacy against a certain kind of attack named differential attack [13]. In our basic scheme, although users' encrypted data are aggregated at the GW so that the individual electricity usage will not be disclosed by the CC or the adversary, their data are still vulnerable to the differential attack which violates users' privacy through analyzing the aggregated data. Specifically, If the adversary acquires the aggregations of two similar data sets $D_1, D_2 \subseteq \mathbb{U}$, where $D_1 = D_2 + U_k$, assume $A(D)$ denotes the sum aggregation on data set D, then $A(D_1) - A(D_2)$ is exactly the data of user U_k.

5.4.1 Differential Privacy

In order to resist the special differential attack, it suffices to enhance our scheme with differential privacy techniques. Differential privacy is first proposed by Dwork in [13] in 2006. By adding appropriately chosen noises, e.g. from Laplace distribution, symmetric geometric distribution, etc., to the aggregation results, the outputs will become indistinguishable for similar inputs (data sets). To the best of our knowledge, all previous literatures consider providing differential privacy over a certain kind of aggregation, i.e. the summation, while our scheme involves multifunctional aggregations, thus we also need to consider differential privacy of these more complex aggregations such as variance and one-way ANOVA. Before showing the details of our enhanced scheme, we first give the formal definition of differential privacy as follows.

Differential Privacy. Let $\epsilon > 0$. The randomized function A is said to give ϵ-differential privacy if for all data sets D_1 and D_2 differing in at most one element, and all $S \subseteq Range(A)$,

$$\Pr[A(D_1) \in S] \leq \exp(\epsilon) \cdot \Pr[A(D_2) \in S]. \tag{5.15}$$

In our enhanced scheme, the noise is chosen from the symmetric geometric distribution Geom(α), $0 < \alpha < 1$, which can be regarded as a discrete approximation of Laplace distribution Lap(λ) (where $\alpha \approx \exp(-\frac{1}{\lambda})$). The use of geometric distribution for the noise was pioneered by Ghosh et al. in [14]. The probability density function of the geometric distribution Geom(α) is

$$\Pr[X = x] = \frac{1-\alpha}{1+\alpha}\alpha^{|x|}. \tag{5.16}$$

Formally, if the sensitivity of aggregation function A(D) is

$$\Delta \mathrm{A} = \max_{D_1, D_2} ||\mathrm{A}(D_1) - \mathrm{A}(D_2)||_1 \tag{5.17}$$

for all data sets D_1 and D_2 differing in at most one element, then by adding geometric noise r randomly chosen from Geom($\exp(-\frac{\epsilon}{\Delta \mathrm{A}})$) to the aggregation data, the perturbed result can achieve ϵ-differential privacy, i.e. for any integer $k \in Range(\mathrm{A})$,

$$\Pr[\mathrm{A}(D_1) + r = k] \leq \exp(\epsilon) \cdot \Pr[\mathrm{A}(D_2) + r = k]. \tag{5.18}$$

5.4.2 *Description of the Enhanced Version*

The *System Initialization* phase and *User Report Generation* phase in our enhanced scheme are exactly the same as those in the basic scheme, thus we start from describing the *Privacy-preserving Report Aggregation* phase.

5.4.2.1 Privacy-Preserving Report Aggregation

In our enhanced scheme, we let the GW perturb encrypted aggregations to ensure differential privacy. Specifically, since the GW can also compute $g_\gamma = H(t_\gamma)$ for the current time point t_γ, after randomly choosing a noise from the geometric distribution, the GW could perturb the aggregation by simply multiplying the encrypted noise to the aggregation. In order to achieve ϵ-differential privacy for a given ϵ, the GW should add a noise proportional to the sensitivity of aggregation. However, different aggregations have different sensitivities, thus the GW needs to carefully choose the parameters of geometric distribution to achieve the desired privacy levels.

- *Average Aggregation:* In the case of average aggregation, the encrypted aggregation $\mathrm{A}_{1,\gamma}$ is the encryption of the sum of all users' data $M_{sum} = \sum_{i=1}^{n} m_{i,\gamma}$. Let $\mathrm{A}_1(D) = \sum_{U_i \in D} m_i$, then for any two data sets D_1 and D_2 differing in at most one element, $|\mathrm{A}_1(D_1) - \mathrm{A}_1(D_2)| \leq W$ holds, thus the sensitivity of A_1 is $\Delta \mathrm{A}_1 = W$.

After computing the aggregation $\mathrm{A}_{1,\gamma}$ as described in *Step-1* of the basic scheme, the GW takes the following steps further in our enhanced scheme.

Step-2: The GW computes the sensitivity of aggregation as $\Delta\mathrm{A}_1 = W$, and then randomly chooses a noise $\tilde{m}_{1,\gamma}$ from the geometric distribution $\mathrm{Geom}(\exp(-\frac{\epsilon}{W}))$.

Step-3: The GW chooses a random number $\tilde{r}_{1,\gamma} \in \mathbb{Z}_N^*$, and computes the final aggregation as

$$\tilde{\mathrm{A}}_{1,\gamma} = \mathrm{A}_{1,\gamma} \cdot g_\gamma^{\tilde{m}_{1,\gamma}} h^{\tilde{r}_{1,\gamma}}. \tag{5.19}$$

Step-4: Eventually, the GW reports the aggregated data $\tilde{\mathrm{A}}_{1,\gamma}$ to the CC for further computation.

- *Variance Aggregation:* In the case of variance aggregation, the encrypted aggregation $\mathrm{A}_{2,\gamma}$ is the encryption of the square of sum of all users' data $M_{sqrsum} = (\sum_{i=1}^n m_{i,\gamma})^2$. Let $\mathrm{A}_2(D) = (\sum_{U_i \in D} m_i)^2$, then $|\mathrm{A}_2(D_1) - \mathrm{A}_2(D_2)| \le (2n-1)W^2$, thus $\Delta\mathrm{A}_2 = (2n-1)W^2$.

$\mathrm{A}_{3,\gamma}$ is the encryption of the sum of squares of all users' data $M_{sumsqr} = \sum_{i=1}^n m_{i,\gamma}^2$. Let $\mathrm{A}_3(D) = \sum_{U_i \in D} m_i^2$, then $|\mathrm{A}_3(D_1) - \mathrm{A}_3(D_2)| \le W^2$, thus $\Delta\mathrm{A}_3 = W^2$.

After computing the aggregations $\mathrm{A}_{2,\gamma}, \mathrm{A}_{3,\gamma}$ as described in *Step-1* and *Step-2* of the basic scheme, the GW continues to take the following steps in our enhanced scheme.

Step-3: The GW computes the sensitivities of aggregations as $\Delta\mathrm{A}_2 = (2n-1)W^2$ and $\Delta\mathrm{A}_3 = W^2$, and then randomly chooses noise $\tilde{m}_{2,\gamma}$ from $\mathrm{Geom}(\exp(-\frac{\epsilon}{(2n-1)W^2}))$ and noise $\tilde{m}_{3,\gamma}$ from $\mathrm{Geom}(\exp(-\frac{\epsilon}{W^2}))$.

Step-4: The GW chooses two random numbers $\tilde{r}_{2,\gamma}, \tilde{r}_{3,\gamma} \in \mathbb{Z}_N^*$, and computes the final aggregations as

$$\tilde{\mathrm{A}}_{2,\gamma} = \mathrm{A}_{2,\gamma} \cdot e(g_\gamma, g_\gamma)^{\tilde{m}_{2,\gamma}} e(g_\gamma, h)^{\tilde{r}_{2,\gamma}}, \tag{5.20}$$

and

$$\tilde{\mathrm{A}}_{3,\gamma} = \mathrm{A}_{3,\gamma} \cdot e(g_\gamma, g_\gamma)^{\tilde{m}_{3,\gamma}} e(g_\gamma, h)^{\tilde{r}_{3,\gamma}}. \tag{5.21}$$

Step-5: Eventually, the GW reports the aggregated data $(\tilde{\mathrm{A}}_{2,\gamma}, \tilde{\mathrm{A}}_{3,\gamma})$ to the CC for further computation.

- *One-Way ANOVA Aggregation:* In the case of one-way ANOVA aggregation, the encrypted aggregation $\mathrm{A}_{4,\gamma}$ is the encryption of the sum of squares of all users' data $M_4 = \sum_{j=1}^s \sum_{i=1}^n m_{ij}^2$. Let $\mathrm{A}_4(D) = \sum_{j=1}^s \sum_{i=1}^n m_{ij}^2$, then $|\mathrm{A}_4(D_1) - \mathrm{A}_4(D_2)| \le W^2$, thus $\Delta\mathrm{A}_4 = W^2$.

$\mathrm{A}_{5,\gamma}$ is the encryption of the sum of squares of sums of data within each pricing strategy $M_5 = \sum_{j=1}^{s}(\sum_{i=1}^{n} m_{ij})^2$. Let $\mathrm{A}_5(D) = \sum_{j=1}^{s}(\sum_{i=1}^{n} m_{ij})^2$, then $|\mathrm{A}_5(D_1) - \mathrm{A}_5(D_2)| \leq (2n-1)W^2$, thus $\Delta\mathrm{A}_5 = (2n-1)W^2$.

$\mathrm{A}_{6,\gamma}$ is the encryption of the square of sum of all data $M_6 = (\sum_{j=1}^{s}\sum_{i=1}^{n} m_{ij})^2$. Let $\mathrm{A}_6(D) = (\sum_{j=1}^{s}\sum_{i=1}^{n} m_{ij})^2$, then $|\mathrm{A}_6(D_1) - \mathrm{A}_6(D_2)| \leq (2ns-1)W^2$, thus $\Delta\mathrm{A}_6 = (2ns-1)W^2$.

After computing the aggregations $\mathrm{A}_{4,\gamma}, \mathrm{A}_{5,\gamma}$ and $\mathrm{A}_{6,\gamma}$ as described in *Step-1*, *Step-2* and *Step-3* of the basic scheme, the GW takes the following steps additionally in our enhanced scheme.

Step-3: The GW computes the sensitivities of aggregations as $\Delta\mathrm{A}_4 = W^2$, $\Delta\mathrm{A}_5 = (2n-1)W^2$ and $\Delta\mathrm{A}_6 = (2ns-1)W^2$, and then randomly chooses noises $\tilde{m}_{4,\gamma}, \tilde{m}_{5,\gamma}$ and $\tilde{m}_{6,\gamma}$ from geometric distribution $\mathrm{Geom}(\exp(-\frac{\epsilon}{W^2}))$, $\mathrm{Geom}(\exp(-\frac{\epsilon}{(2n-1)W^2}))$ and $\mathrm{Geom}(\exp(-\frac{\epsilon}{(2ns-1)W^2}))$ respectively.

Step-4: The GW chooses three random numbers $\tilde{r}_{4,\gamma}, \tilde{r}_{5,\gamma}, \tilde{r}_{6,\gamma} \in \mathbb{Z}_N^*$, and computes the final aggregations as

$$\tilde{\mathrm{A}}_{4,\gamma} = \mathrm{A}_{4,\gamma} \cdot e(g_\gamma, g_\gamma)^{\tilde{m}_{4,\gamma}} e(g_\gamma, h)^{\tilde{r}_{4,\gamma}}, \tag{5.22}$$

$$\tilde{\mathrm{A}}_{5,\gamma} = \mathrm{A}_{5,\gamma} \cdot e(g_\gamma, g_\gamma)^{\tilde{m}_{5,\gamma}} e(g_\gamma, h)^{\tilde{r}_{5,\gamma}}, \tag{5.23}$$

and

$$\tilde{\mathrm{A}}_{6,\gamma} = \mathrm{A}_{6,\gamma} \cdot e(g_\gamma, g_\gamma)^{\tilde{m}_{6,\gamma}} e(g_\gamma, h)^{\tilde{r}_{6,\gamma}}. \tag{5.24}$$

Step-5: Eventually, the GW reports the aggregated data $(\tilde{\mathrm{A}}_{4,\gamma}, \tilde{\mathrm{A}}_{5,\gamma}, \tilde{\mathrm{A}}_{6,\gamma})$ to the CC for further computation.

5.4.2.2 Secure Report Reading

The secure report reading phase in the enhanced scheme is almost the same as in the basic scheme. The only difference is the outputs that CC gets are no longer exact ones, but the aggregations with geometric noises which achieve ϵ-differential privacy. Therefore, even though the adversary $\mathcal{A}$ has intruded into the database of CC, $\mathcal{A}$ still cannot launch the differential attacks to disclose individual user's data.

5.5 Security and Utility Analysis

In this section, we discuss the security issues involved in the basic scheme and the enhanced scheme, specifically, to protect the users' electricity usage privacy against a powerful adversary $\mathcal{A}$. In addition, we also analyze the utility of the enhanced scheme in terms of relative error involved by the noise.

5.5.1 Security Analysis

As mentioned in our security model, we consider a powerful adversary $\mathscr{A}$ who can not only eavesdrop the communication flows in the system, but also intrude into the databases of GW and CC to disclose the stored data. More seriously, the adversary $\mathscr{A}$ can also launch differential attack to acquire individual user's data by analyzing the aggregated data. Since the main purpose of the adversary $\mathscr{A}$ is to threaten users' privacy, we do not consider tampering in this work, although it could be avoided by involving some verification techniques. Below we show that our proposed schemes can effectively resist various attacks launched by the adversary $\mathscr{A}$.

- *The users' electricity usage privacy is protected in our MuDA scheme.* In order to violate users' data privacy, the adversary $\mathscr{A}$ could reside in the residential area to eavesdrop the reports transmitted from users to the GW. Suppose the adversary $\mathscr{A}$ has eavesdropped a report of user U_i at time point t_γ, i.e. $C_{i,\gamma} = g_\gamma^{m_{i,\gamma}} \cdot h^{r_{i,\gamma}}$. Since the household electricity usage $m_{i,\gamma}$ within 15 min is probably a small value, the adversary $\mathscr{A}$ may try to launch a brute-force attack by exhaustedly testing each possible value of $m_{i,\gamma}$. However, since the BGN PKE is semantic secure against the chosen ciphertext attack, the adversary $\mathscr{A}$ is not able to recover U_i's usage data without knowing the random number $r_{i,\gamma}$. Therefore, even if the adversary $\mathscr{A}$ eavesdrops $C_{i,\gamma}$, he still cannot identify the corresponding content. Moreover, after receiving users' reports, the GW does not decrypt them but aggregate them directly, thus even though the adversary $\mathscr{A}$ intrudes into the database of GW to steal the stored data, he still cannot acquire users' electricity usage data. In addition, the adversary could also intrude into the database of the CC, but after decryption, the outputs CC gets are all aggregations of users' data which do not leak individual user's data at all. Therefore, from the above three aspects, the individual user's report is privacy-preserving in our proposed MuDA schemes.
- *The users' electricity usage differential privacy is achieved in our enhanced MuDA scheme.* Although as abovementioned the adversary $\mathscr{A}$ cannot acquire individual user's data through intruding into the CC's database, the adversary $\mathscr{A}$ still can launch some differential attacks to threaten users' privacy if he gets aggregations of two adjacent data sets. In our enhanced MuDA scheme, for a given privacy level ϵ, the GW perturbs the aggregations without decryption by adding appropriate geometric noises in the form of ciphertext. In this way, the aggregation results all achieve ϵ-differential privacy. For example, in the one-way ANOVA aggregation, the GW adds noise $\tilde{m}_4$, which is chosen from $\text{Geom}(\exp(-\frac{\epsilon}{W^2}))$, to the exact aggregation A_4 to get the perturbed aggregation $\tilde{\text{A}}_4$. Assume the adversary $\mathscr{A}$ acquires two perturbed aggregations $u + \tilde{m}_4^{(u)}$ and $v + \tilde{m}_4^{(v)}$, where u and v are two adjacent aggregations while $\tilde{m}_4^{(u)}$ and $\tilde{m}_4^{(v)}$ are the corresponding geometric noises. Since $u - v \leq W^2$, for any integer k,

$$\begin{aligned}\eta =&\Pr[u+\tilde{m}_4^{(u)}=k]/\Pr[v+\tilde{m}_4^{(v)}=k]\\ =&\Pr[\tilde{m}_4^{(u)}=k-u]/\Pr[\tilde{m}_4^{(v)}=k-v]\\ =&\left(\frac{1-\alpha}{1+\alpha}\alpha^{|k-u|}\right)/\left(\frac{1-\alpha}{1+\alpha}\alpha^{|k-v|}\right)\\ =&\alpha^{|k-u|-|k-v|},\end{aligned} \tag{5.25}$$

Since $-|u-v| \le |k-u| - |k-v| \le |u-v|$ and $0 < \alpha < 1$,

$$\begin{aligned}\alpha^{|u-v|} \le& \eta \le \alpha^{-|u-v|}\\ \alpha^{W^2} \le \alpha^{|u-v|} \le& \eta \le \alpha^{-|u-v|} \le \alpha^{-W^2}\\ (e^{-\frac{\epsilon}{W^2}})^{W^2} \le& \eta \le (e^{-\frac{\epsilon}{W^2}})^{-W^2}\\ e^{-\epsilon} \le& \eta \le e^{\epsilon}\end{aligned} \tag{5.26}$$

Thus $\tilde{\text{A}}_4$ satisfy ϵ-differential privacy; similarly, the other aggregations mentioned in our enhanced MuDA scheme also achieve ϵ-differential privacy. Therefore, even though the adversary $\mathscr{A}$ obtains aggregations of two adjacent data sets, the difference of the two aggregation results still does not leak the individual user's private data to $\mathscr{A}$ at all.

5.5.2 Utility Analysis

In order to achieve differential privacy, an appropriate magnitude of noise $\tilde{m}_i$ is added to the exact aggregation A_i, which obscures individual user's data but also introduces error into the perturbed aggregation $\tilde{\text{A}}_i$. One most commonly used error measurement is the Mean Absolute Deviation, which is the expectation $\mathbb{E}|\tilde{\text{A}}_i - \text{A}_i| = \mathbb{E}|\tilde{m}_i|$. However, as the sensitivity ΔA_i increases, although the noise $|\tilde{m}_i|$ becomes larger, the exact aggregation A_i also increases accordingly. Thus in order to better investigate the added noise compared to the exact aggregation, we use the Relative Error $\zeta_i = \frac{|\tilde{\text{A}}_i - \text{A}_i|}{\text{A}_i}$ as the measurement of utility. Therefore, for a given A_i, the expectation of ζ_i is

$$\begin{aligned}\mathbb{E}(\zeta_i) =& \frac{\mathbb{E}|\tilde{\text{A}}_i - \text{A}_i|}{\text{A}_i}\\ =& \frac{\mathbb{E}|\tilde{m}_i|}{\text{A}_i},\end{aligned} \tag{5.27}$$

where

$$
\begin{aligned}
\mathbb{E}|\tilde{m}_i| &= \sum_{x=-\infty}^{\infty} |x| \cdot \Pr[X = x] \\
&= \sum_{x=-\infty}^{\infty} |x| \cdot \frac{1-\alpha}{1+\alpha} \alpha^{|x|} \\
&= \frac{2}{1+\alpha} \cdot \sum_{x=1}^{\infty} x(1-\alpha) \cdot \alpha^{x} \\
&= \frac{2}{1+\alpha} \cdot \left(\sum_{x=1}^{\infty} x \cdot \alpha^{x} - \sum_{x=1}^{\infty} x \cdot \alpha^{x+1}\right) \\
&= \frac{2}{1+\alpha} \cdot \sum_{x=1}^{\infty} \alpha^{x} \\
&= \frac{2}{1+\alpha} \cdot \frac{\alpha}{1-\alpha} \qquad (\because\ 0 < \alpha < 1) \\
&= \frac{2\alpha}{1-\alpha^2}.
\end{aligned} \tag{5.28}
$$

Since $\alpha = \exp(-\frac{\epsilon}{\Delta \mathrm{A}_i})$, then $\mathbb{E}|\tilde{m}_i| = 2e^{-\frac{\epsilon}{\Delta \mathrm{A}_i}}/(1 - e^{-\frac{2\epsilon}{\Delta \mathrm{A}_i}})$. According to the Taylor series, for any real number $z \approx 0$, $e^{-z} \approx 1 - z$. Let $z = \frac{\epsilon}{\Delta \mathrm{A}_i}$, we have $\mathbb{E}|\tilde{m}_i| \approx 2(1-z)/(1-(1-2z)) = \frac{1-z}{z} = \frac{\Delta \mathrm{A}_i - \epsilon}{\epsilon}$. Therefore, $\mathbb{E}(\zeta_i) = \frac{\mathbb{E}|\tilde{m}_i|}{\mathrm{A}_i} = \frac{\Delta \mathrm{A}_i - \epsilon}{\epsilon \mathrm{A}_i}$.

Figure 5.3 shows the relative error of A_4 with different numbers of users as an example. In the simulation, we assume that each meter's readings is in the range from $\{0.0, 0.1, \cdots, 5.0\}$ and $s = 5$. We vary the number of users from 3000 to 15,000 and compare the utility of our protocol under the privacy parameter $\epsilon = 0.1$. Simulation result shows that our enhanced protocol produces little noise with different numbers of users n.

5.6 Performance Evaluation

In this section, we evaluate the performance of the proposed MuDA scheme in terms of computation complexity and communication overhead in the smart grid communication. We compare our proposed scheme with the one proposed by Shi et al. [11] which also supports privacy-preserving aggregation and differential privacy. The basic construction of [11] achieves privacy-preserving aggregation by assigning each user a private key and making all these keys sum up to zero. Similarly, the decryption of their basic construction also needs the aggregator to compute the discrete logarithm in a cyclic group. Their enhanced version also preserves differential privacy, however, the noise added to the aggregation is not generated by the aggregator but by users in a distributed way. Thus their scheme can

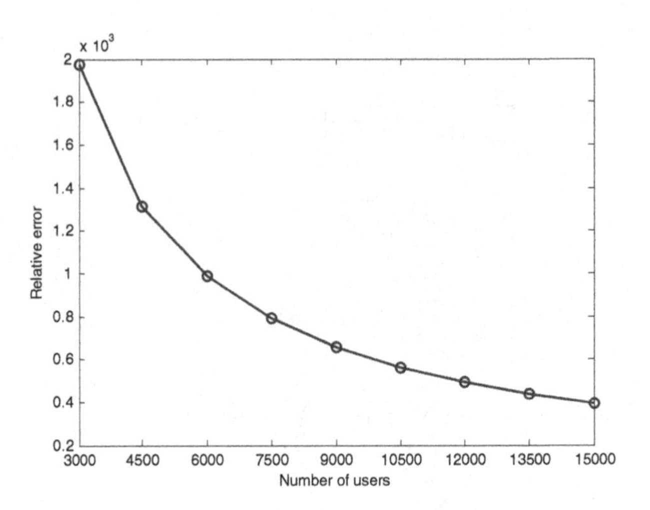

Fig. 5.3 Relative error of A_4

be used in the scenario that the aggregator is not trustable, but at the same time, the distributed noises also decrease the utility of the aggregation. Actually, our scheme can not only support privacy-preserving aggregation and differential privacy, but also provide multifunctional aggregations. Before comparing computation and communication overhead of the two schemes, we first optimize our basic scheme to reduce the computation cost of CC, while introducing a little more communication overhead only.

5.6.1 Computational Optimization of Basic Scheme

In the *Privacy-preserving Report Aggregation* phase of our basic scheme, the GW has computed one kind of aggregation named "square of sum", such as $A_{2,\gamma}, A_{5,\gamma}$ and $A_{6,\gamma}$. These aggregations will cost the CC expect time $O(nW), O(\sqrt{s}nW)$ and $O(nsW)$ respectively for decryption. In fact, we can modify our basic scheme to let the GW compute the "sum" only and leave the calculation of "square" to the CC. For example, when computing $A_{2,\gamma}$, the GW computes

$$\begin{aligned} A'_{2,\gamma} &= \prod_{i=1}^{n} C_{i,\gamma} \\ &= \prod_{i=1}^{n} (g_\gamma^{m_{i,\gamma}} \cdot h^{r_{i,\gamma}}) \\ &= g_\gamma^{\sum_{i=1}^{n} m_{i,\gamma}} \cdot h^{\sum_{i=1}^{n} r_{i,\gamma}}, \end{aligned} \tag{5.29}$$

which is actually the encryption of sum of all users' data. After receiving $A'_{2,\gamma}$, the CC decrypts it and obtains the sum of users' data M_{sum} and computes the variance as

$$M_{var} = \frac{1}{n}M_{sumsqr} - (\frac{1}{n}M_{sum})^2,$$

Note that before optimization, the computation cost of CC is $O(nW) + O(\sqrt{n}W) = O(nW)$. After modification, the computation cost becomes $O(\sqrt{nW}) + O(\sqrt{n}W) = O(\sqrt{n}W)$ which is an obvious improvement especially when n is large. And the communication cost is still two aggregations $(A'_{2,\gamma}, A_{3,\gamma})$ which is the same as before.

Similarly, the computation cost of $A_{5,\gamma}$ is reduced from $O(\sqrt{s}nW)$ to $O(s\sqrt{nW})$, but the communication cost will increase from one aggregation $(A_{5,\gamma})$ to s aggregations $(A^{(1)}_{5,\gamma}, A^{(2)}_{5,\gamma}, \cdots, A^{(s)}_{5,\gamma})$ which are the sums of users' data within each pricing strategy respectively. And the computation cost of $A_{6,\gamma}$ is reduced from $O(nsW)$ to $O(\sqrt{nsW})$ while the communication cost is unchanged. As a result, the computation cost of CC to calculate the one-way ANOVA is reduced from $O(\sqrt{ns}W) + O(\sqrt{s}nW) + O(nsW) = O(nsW)$ to $O(\sqrt{ns}W) + O(s\sqrt{nW}) + O(\sqrt{nsW}) = O(c\sqrt{n})$, where $c = \max(W\sqrt{s}, s\sqrt{W})$ is a small constant since s and W are both small values.

5.6.2 Computation Complexity

Above we have already discussed the computation complexity of the CC in terms of computing group based discrete logarithm. Now we focus on the computation cost of the GW in the *Privacy-preserving Report Aggregation* phase. In our basic scheme, to compute the aggregation of users' data, the GW needs to perform two primary operations, the bilinear pairing and the group based multiplication. Let T_p denote the computation time of bilinear pairing and T_m denote the time of group based multiplication. Then the total computation time of average aggregation, i.e. computing $A_{1,\gamma}$, is obviously $(n-1) \cdot T_m$. Since computing $A_{2,\gamma}$ needs n pairings and $n-1$ multiplications and computing $A_{3,\gamma}$ needs $n-1$ multiplications, the total computation time of variance aggregation (after optimization) is $n \cdot T_p + 2(n-1) \cdot T_m$. Similarly, computing $A_{4,\gamma}$ needs ns pairings and $ns-1$ multiplications, computing $A^{(j)}_{5,\gamma}$ $(j = 1, 2, \ldots, s)$ needs $n-1$ multiplications each, and computing $A_{6,\gamma}$ needs $ns-1$ multiplications. Thus the total computation time of one-way ANOVA aggregation (after optimization) is $ns \cdot T_p + (3ns - s - 2) \cdot T_m$.

In the enhanced version of our scheme, the GW needs to generate a noise and add it to each aggregation, which introduces an encryption and another multiplication. However, the additional computation cost is regardless of n, the number of users, it could hardly influence the computation complexity of GW's operations.

Shi et al.'s scheme can only support the sum aggregation, thus we only compare the computation cost of average aggregation with it. In Shi et al.'s scheme, the aggregator computes the aggregation of users' data by multiplying them together, which also needs $n-1$ group based multiplications. To decrypt the aggregation, the aggregator computes the group based discrete logarithm with expected time $\sqrt{nW}$ which is also the same as our scheme. Thus the computation cost of our scheme is intuitively the same as Shi et al.'s.

5.6.3 Communication Overhead Comparison

We consider the communication overhead of the proposed MuDA scheme in two metrics, the individual user communication cost and the overall communication cost. In our proposed schemes, no matter the basic scheme, optimized basic scheme or enhanced version, no matter what kind of aggregation is chosen, the individual user U_i needs to transmit only one ciphertext $C_{i,\gamma}$ to the GW. Thus the size of U_i's report at a time point is $|C_{i,\gamma}| = 1024$ bits if we choose the security parameter $\tau = 512$. Note that we do not consider other payload such as user ID and timestamp which are relatively short compared with the report. Let $L_N = 1024$ bits denote the length of $N = pq$, then the communication overhead of individual user in our scheme is always L_N at every time point. Since Shi et al.'s scheme can only support sum aggregation, in order to compute the variance, their scheme has to require the users to additionally transmit another report $C'_{i,\gamma} = g_\gamma^{\sum_{i=1}^{n} m_{i,\gamma}^2} \cdot h^{\sum_{i=1}^{n} r'_{i,\gamma}}$, then the aggregator could compute the variance by aggregating $C_{i,\gamma}$ and $C'_{i,\gamma}$ respectively. Thus the communication overhead of each user is $2 \cdot L_N$ at a time point. Similarly, if one-way ANOVA aggregation is required, Shi et al.'s scheme also needs users to submit $C'_{i,\gamma}$ for computation, since the aggregator could not compute squares of users' data by itself. Thus the communication overhead of the individual user is $2 \cdot L_N$, too. Because the two schemes have the same user communication cost in the average aggregation, we only plot the individual user communication overhead of variance aggregation and one-way ANOVA aggregation, as shown in Fig. 5.4.

Next, we consider the overall communication of both MuDA and Shi et al.'s scheme. In the average aggregation of MuDA, the GW collects n users' reports and aggregates them into one report $A_{1,\gamma}$ whose size is just L_N, then transmits $A_{1,\gamma}$ to the CC. Thus by summing up all users' communication overhead and the cost from GW to CC, the total communication overhead is $(n+1) \cdot L_N$. Similarly, the communication cost from the GW to CC are $2 \cdot L_N$ and $(s+2) \cdot L_N$ in the optimized variance aggregation and one-way ANOVA aggregation, respectively. Therefore the overall communication overhead of variance aggregation in our MuDA scheme is $(n+2) \cdot L_N$ and that of one-way ANOVA aggregation is $(n+s+2) \cdot L_N$. Since Shi et al.'s scheme does not have a CC in their system model, the overall communication cost is just the sum of all users' communication overhead. Therefore, the overall communication overhead of average aggregation, variance aggregation and one-way

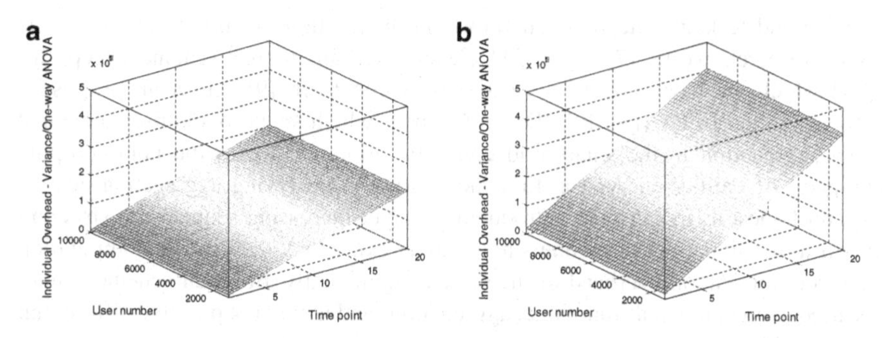

Fig. 5.4 Communication overhead of individual user. (**a**) Individual communication overhead of MuDA. (**b**) Individual communication overhead of Shi et al.'s scheme

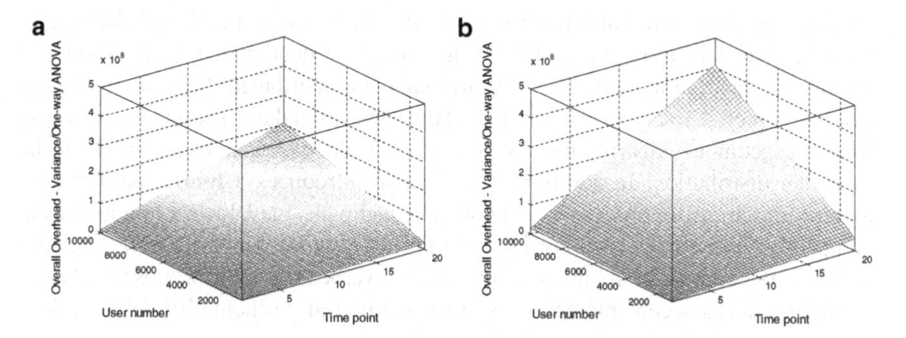

Fig. 5.5 Overall communication overhead of (**a**) MuDA and (**b**) Shi et al.'s scheme

ANOVA aggregation are $n{\cdot}L_N$, $2n{\cdot}L_N$ and $2n{\cdot}L_N$, respectively. Because two schemes have the same overall communication cost in the average aggregation, we further plot in Fig. 5.5 the overall communication overhead of variance aggregation and one-way ANOVA aggregation (we choose $s = 5$, since $n + 2 \approx n + s + 2$ when n is large, we ignore the difference) in terms of user number n and running time γ. It is shown that the MuDA scheme significantly reduces the overall communication overhead in the smart grid communication.

From the above analysis, the proposed MuDA scheme is indeed efficient in terms of communication cost while providing multifunctional aggregations. Since at each reporting time point, all n users will report their packets nearly at the same time, it is obvious that our MuDA scheme can significantly reduce the packet transmitting time, especially when the user number n is large.

5.7 Related Work

In smart grid communication, using aggregation techniques to collect statistical data can significantly decrease computation and communication overhead of the CC, assisting the CC to provide efficient and low-latency services such as forecasting, pricing and leakage detection. Therefore, in this section, we briefly discuss some other research works [7, 9, 12, 21] closely relevant to our scheme except for [11] which we have already discussed in Sect. 5.6. In [9], Lu et al. propose a scheme called EPPA (i.e., PPMDA in Chap. 3) which supports multi-dimensional data aggregation in the smart grid communication. It utilizes the homomorphic property of Paillier encryption to achieve privacy-preserving aggregation against a semi-honest aggregator. By introducing a super-increasing sequence to structure multi-dimensional data into one single ciphertext, EPPA reduces the computation and communication overhead of the system significantly. But their scheme cannot be used to compute multifunctional aggregations and it does not preserve differential privacy either.

In [21], Rastogi et al. propose a differentially private aggregation scheme named PASTE which also uses Paillier encryption to achieve privacy-preserving aggregation. But different from Lu et al. [9], they use a distributed decryption algorithm so that users can assist the aggregator to decrypt the aggregation and remove the random numbers. PASTE also supports distributed differential privacy, just like Shi et al.'s scheme does in [11]. However, instead of picking random noise from the geometric distribution, users in PASTE choose random noises from the Gaussian distribution. In addition, PASTE also introduces a fourier perturbation algorithm to transmit time series data of users and picks the k lowest frequencies to compactly represent the high-level trends of users' data. This algorithm significantly deceases the computation and communication overhead of their proposed scheme, while introduces acceptable errors. Similar as Lu et al.'s scheme [9], PASTE does not provide multifunctional aggregations, either.

Jia et al. also propose a privacy-preserving aggregation scheme in smart grid communication in [7]. In their scheme, users divide their readings into shares, and use these shares as the coefficients of a polynomial. Only the aggregator with private key can recover the polynomials and acquire the sum of shares, while individual user's readings are protected. In addition, they take the human-factor into consideration so that their scheme also preserves differential privacy. In [12], Chen et al. propose a privacy-preserving data aggregation scheme with fault tolerance for smart grid communications. They consider protecting users' data privacy under a strong adversary model, where the adversary can even compromise a few servers at the control center. In addition, their proposed scheme is fault tolerant, i.e. their system can work normally even if some smart meters or some servers at the control center occur malfunction. However, neither Jia et al.'s scheme nor Chen et al.'s scheme could support multifunctional aggregations.

Although our proposed MuDA scheme addresses the same issue as the above literatures to provide efficient and privacy-preserving aggregation in smart grid communications, our research focuses are different: (a) in our MuDA scheme, the GW can compute multiple aggregations based on the requirement of CC without requesting the users to transmit any other reports; and (b) except for average aggregation, our MuDA scheme can also achieve differential privacy for other more complex aggregations such as variance aggregation and one-way ANOVA aggregation.

5.8 Summary

In this chapter, we have proposed a privacy-preserving aggregation scheme MuDA that supports multifunctional aggregations, i.e. average aggregation, variance aggregation and one-way ANOVA aggregation. We protect the users' individual electricity usage privacy against an adversary that is curious about user activity patterns. Detailed security analysis and utility analysis show that the proposed MuDA scheme can prevent the adversary from disclosing individual users' data even if he launches the differential attacks, while just introducing acceptable noise. In addition, through extensive performance evaluation, we have also demonstrated that the proposed scheme is communication-efficient compared with relevant schemes.

References

1. L. Chen, R. Lu, Z. Cao, K. Alharbi, and X. Lin, "Muda: Multifunctional data aggregation in privacy-preserving smart grid communications," *Peer-to-Peer Networking and Applications*, vol. 8, no. 5, pp. 777–792, 2015. [Online]. Available: http://dx.doi.org/10.1007/s12083-014-0292-0
2. G. Maas, M. Bial, and J. Fijalkowski, "Final report-system disturbance on 4 november 2006," *Union for the Coordination of Transmission of Electricity in Europe, Tech. Rep*, 2007.
3. G. Andersson, P. Donalek, R. Farmer, N. Hatziargyriou, I. Kamwa, P. Kundur, N. Martins, J. Paserba, P. Pourbeik, J. Sanchez-Gasca *et al.*, "Causes of the 2003 major grid blackouts in North America and Europe, and recommended means to improve system dynamic performance," *IEEE Transactions on Power Systems*, vol. 20, no. 4, pp. 1922–1928, 2005.
4. M. Hashmi, S. Hänninen, and K. Mäki, "Survey of smart grid concepts, architectures, and technological demonstrations worldwide," in *Innovative Smart Grid Technologies (ISGT Latin America), 2011 IEEE PES Conference on*. IEEE, 2011, pp. 1–7.
5. H. Li, X. Lin, H. Yang, X. Liang, R. Lu, and X. Shen, "EPPDR: an efficient privacy-preserving demand response scheme with adaptive key evolution in smart grid," *IEEE Trans. Parallel Distrib. Syst.*, vol. 25, no. 8, pp. 2053–2064, 2014. [Online]. Available: http://doi.ieeecomputersociety.org/10.1109/TPDS.2013.124
6. W. Meng, R. Ma, and H. Chen, "Smart grid neighborhood area networks: a survey," *IEEE Network*, vol. 28, no. 1, pp. 24–32, 2014. [Online]. Available: http://dx.doi.org/10.1109/MNET.2014.6724103
7. W. Jia, H. Zhu, Z. Cao, X. Dong, and C. Xiao, "Human-factor-aware privacy preserving aggregation in smart grid," *IEEE System Journal*, to appear.
8. F. Li, B. Luo, and P. Liu, "Secure information aggregation for smart grids using homomorphic encryption," in *SmartGridComm*, 2010, pp. 327–332.
9. R. Lu, X. Liang, X. Li, X. Lin, and X. Shen, "Eppa: An efficient and privacy-preserving aggregation scheme for secure smart grid communications," *IEEE Trans. Parallel Distrib. Syst.*, vol. 23, no. 9, pp. 1621–1631, 2012.
10. K. Alharbi and X. Lin, "Lpda: A lightweight privacy-preserving data aggregation scheme for smart grid," in *WCSP*, 2012, pp. 1–6.
11. E. Shi, T.-H. H. Chan, E. G. Rieffel, R. Chow, and D. Song, "Privacy-preserving aggregation of time-series data," in *NDSS*, 2011.

12. L. Chen, R. Lu, and Z. Cao, "PDAFT: A privacy-preserving data aggregation scheme with fault tolerance for smart grid communications," *Peer-to-Peer Networking and Applications*, vol. 8, no. 6, pp. 1122–1132, 2015. [Online]. Available: http://dx.doi.org/10.1007/s12083-014-0255-5
13. C. Dwork, "Differential privacy," in *Automata, Languages and Programming, 33rd International Colloquium, ICALP 2006, Venice, Italy, July 10–14, 2006, Proceedings, Part II*, 2006, pp. 1–12. [Online]. Available: http://dx.doi.org/10.1007/11787006_1
14. A. Ghosh, T. Roughgarden, and M. Sundararajan, "Universally utility-maximizing privacy mechanisms," *SIAM J. Comput.*, vol. 41, no. 6, pp. 1673–1693, 2012. [Online]. Available: http://dx.doi.org/10.1137/09076828X
15. C. Dwork, "Differential privacy: A survey of results," in *Theory and Applications of Models of Computation, 5th International Conference, TAMC 2008, Xi'an, China, April 25–29, 2008. Proceedings*, 2008, pp. 1–19. [Online]. Available: http://dx.doi.org/10.1007/978-3-540-79228-4_1
16. R. A. Fisher, *Statistical methods for research workers*. Genesis Publishing Pvt Ltd, 1925.
17. "Wireless medium access control (mac) and physical layer (phy) specifications for low-rate wireless personal area networks (wpans) amendment 4: Physical layer specifications for low data rate wireless smart metering utility networks," *IEEE Std. P802.15.4g/D4 part 15.4*, Apr. 2011.
18. X. Liang, X. Li, R. Lu, X. Lin, and X. Shen, "Udp: Usage-based dynamic pricing with privacy preservation for smart grid," *IEEE Trans. Smart Grid*, vol. 4, no. 1, pp. 141–150, 2013.
19. K. Kursawe, G. Danezis, and M. Kohlweiss, "Privacy-friendly aggregation for the smart-grid," in *PETS*, 2011, pp. 175–191.
20. A. J. Menezes, P. C. Van Oorschot, and S. A. Vanstone, *Handbook of applied cryptography*. CRC press, 1997.
21. V. Rastogi and S. Nath, "Differentially private aggregation of distributed time-series with transformation and encryption," in *SIGMOD*, 2010, pp. 735–746.

Chapter 6
Privacy-Preserving Data Aggregation with Fault Tolerance

In this chapter, we introduce a privacy-preserving data aggregation scheme with fault tolerance, named PDAFT, for smart grid communications [1]. Because PDAFT supports the fault-tolerant feature, even when some user failure or server malfunction occurs, PDAFT can still work well.

6.1 Introduction

Compared with traditional power grid, smart grid has introduced new concepts and offered promising solutions for intelligent electricity generation, transmission, distribution and utilization. By deploying various sensors along with the two-way flows of electricity and communication, a huge amount of real-time information is collected and reported to the control center (CC) for timely monitoring and analyzing the health of power grid, as illustrated in Fig 6.1. Specifically, all the intelligent electric appliances in the residential user's home are connected to a key element, *smart meter*, which periodically records the power consumption of appliances and reports the metering data to a local area *gateway*. Then the gateway collects and forwards data to the control center for further analysis and processing, e.g., making real-time power pricing decisions [2] and detecting power fraud/leakage [3]. In both cases, the control center needs to compute a certain kind of statistics, i.e., summation, of users' data.

There are usually two ways of collecting data in smart grid, one is at a low frequency and the other is at a high frequency [4]. The low-frequency data contains a summary for some periodic power usage, brief enough to avoid privacy leakage. The high-frequency data, e.g., those collected every 15 min, include specific power usage patterns for fine-grained real-time control and optimization. As they are related to users' private lives, the high-frequency data has to be protected from utilities [5, 6].

R. Lu, *Privacy-Enhancing Aggregation Techniques for Smart Grid Communications*, Wireless Networks, DOI 10.1007/978-3-319-32899-7_6

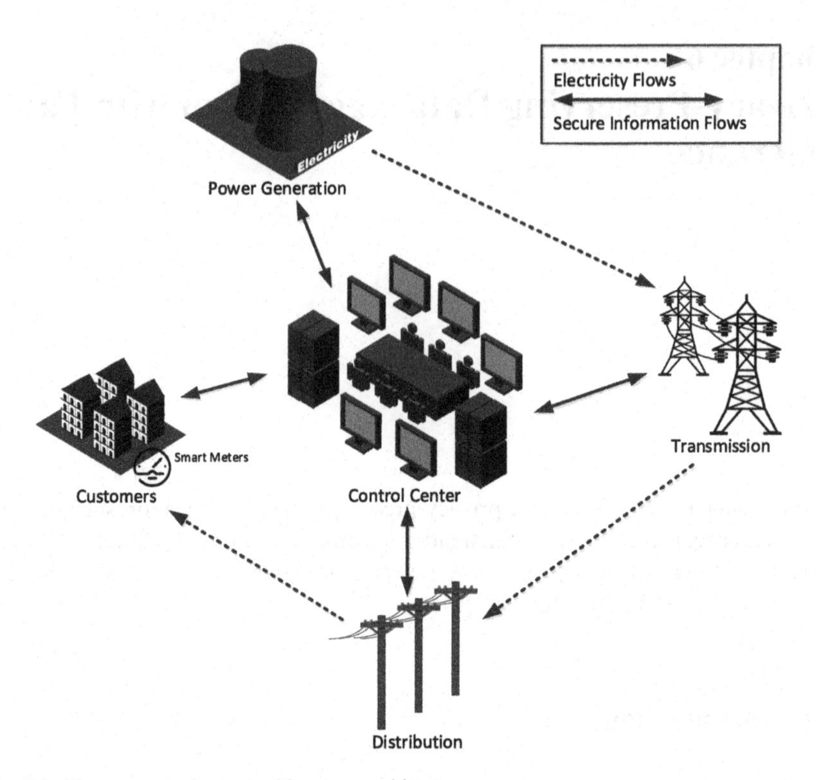

Fig. 6.1 The conceptual smart grid system architecture

To protect user privacy, local gateways should not be able to access the content of consumers' data. By enabling them to aggregate the data, homomorphic encryption techniques [7] may be applied to encrypt users' data. They allow the local gateways to perform summation based aggregation on users' data without decrypting them. Existing data aggregation schemes [7, 8] apply different homomorphic encryptions to achieve the same purpose; however, they only consider protection of user privacy against the gateway (aggregator), while the control center is still free to learn individual users' data. This is because private keys the control center possesses can not only be used to decrypt aggregated data, but also be used to reveal each user's electricity usage. It may also conflict residential users' privacy concerns, especially when the control center is vulnerable to some strong adversaries, i.e. the adversary that can compromise a few servers at the control center to obtain the private keys. Consider a curious control center or a strong adversary that aims to spy on user privacy, these privacy-preserving data aggregation schemes are not sturdy enough to keep user activities unexposed.

Other aggregation schemes such as [9] rely on the following key idea: each user would incorporate a random value into their ciphertext; and the aggregator also incorporates a random value. All of these random values sum up to 0, and would cancel out in the decryption step, such that the aggregator can recover the sum of all users' data, but learn nothing else. One major drawback of these existing works [9–11] is that these schemes are not tolerant of user failures. Even if a single user fails to respond in a certain aggregation round, the servers would not be able to learn anything. This can be a big concern in real-world applications where failures may be unavoidable. On the other hand, although servers at the control center are reliable and robust, they may also sometimes suffer from malfunction or shutdown initiatively to avoid certain attacks. Even worse, some of the servers may be compromised by a strong adversary, thus the servers should also be fault-tolerant or the whole system will fall into paralyzed.

To prevent a strong adversary threatening user privacy and to deal with malfunction of both users and servers, we propose a novel *P*rivacy-preserving *D*ata *A*ggregation with *F*ault *T*olerance (PDAFT) scheme in this chapter. This scheme provides privacy-preserving aggregation against a strong adversary which may compromise a few servers at the control center and supports fault tolerance of smart grid users and servers. As a result, data can be confidentially reported to smart grid control center for real-time monitoring with high reliability. The main contributions of this paper are three-fold.

- Firstly, since the high-frequency user data contains specific power usage patterns, to preserve user privacy, we present a novel PDAFT scheme that protects user electricity usage privacy against a strong adversary which can even compromise a few servers at the control center. Compared with earlier aggregation schemes [7, 8] that only prevent the gateway to reveal user privacy, our proposed scheme leads to much stronger privacy preservation.
- Secondly, inspired by the fact that user and server malfunctions are unavoidable in reality, our proposed PDAFT scheme is designed to support fault tolerance for both user and server failure. Compared with existing works [9–11], our proposed scheme is more reliability in case of user or server malfunctions occur.
- Thirdly, we extend our scheme PDAFT to support temporal aggregation and dynamic users. And through comparative performance analysis, we demonstrate that PDAFT is more efficient than a close aggregation scheme [12] in terms of communication overhead.

The remainder of this chapter is organized as follows. In Sect. 6.2, we introduce the system model, adversary model, security requirements and design goal. Then, we present our detailed PDAFT scheme in Sect. 6.3, followed by its security analysis and performance evaluation in Sects. 6.4 and 6.5, respectively. We also discuss the related work in Sect. 6.6. Finally, we draw our conclusions in Sect. 6.7.

6.2 Problem Formalization

In this section, we formalize our research problems in smart grid communications, including system model, adversary model, security requirements, and design goal.

6.2.1 System Model

As residential users care about their privacy when reporting their electricity usage data to the control center (CC) in smart grid communications, in this work, we mainly focus on how to report users' data to the control center (CC) in a privacy-preserving and fault-tolerant way. Specifically, in our system model, we consider a typical smart grid communication architecture for residential users, which includes a trusted authority (TA), a set of servers $\mathbb{S} = \{S_1, S_2, \ldots, S_k\}$ on behalf of the control center, a residential gateway (GW), and a large number of residential users $\mathbb{U} = \{U_1, U_2, \cdots, U_n\}$ in a residential area (RA), as shown in Fig. 6.2.

Trusted Authority (TA): The TA is a trustable and powerful entity in charge of management of the whole system, for example, initializing the system, registering residential users $\mathbb{U}$ with key materials, and distributing keys for servers $\mathbb{S}$. In general, after initializing the system, the TA will be off-line, i.e., it won't directly participate in the user data reporting unless some exceptions occur in reporting.

Servers ($\mathbb{S} = \{S_1, S_2, \ldots, S_k\}$) at CC: The CC is comprised of a set of servers $\mathbb{S} = \{S_1, S_2, \ldots, S_k\}$, $k \geq 3$, which run synergistically to collect, process and analyze the nearly real-time data for providing reliable services for electrical grid, e.g., real-time monitoring the RA's usage for leakage detecting, fraud detecting, and forecasting [3]. Unlike the TA, servers $\mathbb{S} = \{S_1, S_2, \ldots, S_k\}$ are powerful entities in our system, but some of them could be compromised or paralyzed by a strong adversary. As $\mathbb{S}$ are powerful entities, it will take huge costs for an

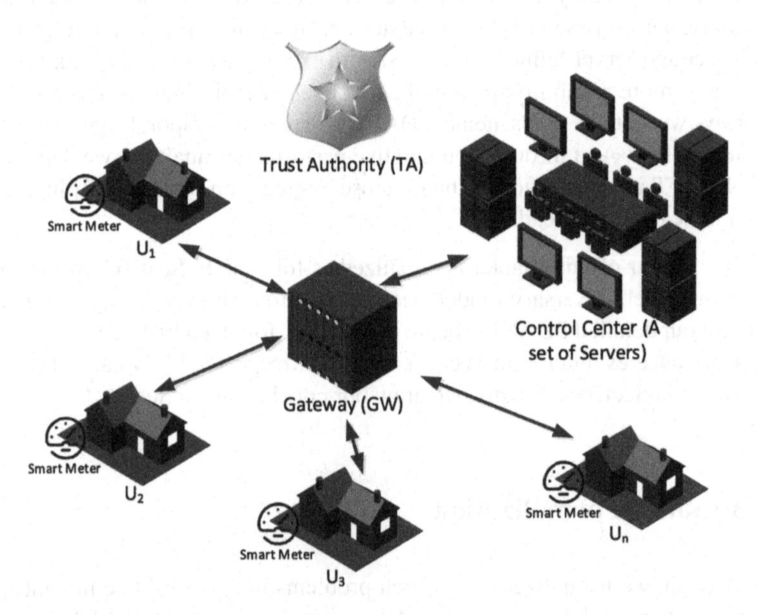

Fig. 6.2 System model under consideration

adversary to compromise even a single server. As a result, it is reasonable for us to assume that the adversary can only compromise a limited number, i.e. no more than $d = \lceil k/2 \rceil - 1$, of servers. In other words, the adversary can only compromise minority of the servers $\mathbb{S} = \{S_1, S_2, \ldots, S_k\}$. Unless other specification, we consider servers $\mathbb{S}$ as the representation of CC in the rest of this paper.

Gateway (GW): The GW is a powerful entity, which connects the CC and residential users, i.e., helping the CC collecting the residential users' nearly real-time usage data. As suggested in the Standards [13], the communication between residential users and the GW is through relatively inexpensive WiFi technology, while the communication between the CC and the GW is through either wired links or any other links with high bandwidth and low delay.

Residential Users $\mathbb{U} = \{U_1, U_2, \cdots, U_n\}$: Each residential user $U_i \in \mathbb{U}$ is equipped with a smart meter and various smart appliances to form a Home Area Network (HAN), which can electronically record the real-time electricity usage data, and report to the CC via the GW in a certain period, i.e., every 15 min. As smart meter is not as powerful as the GW, some meters could be malfunctioning occasionally, i.e., they could stop reporting for a while and will be reset in a late time. However, malfunction of smart meter can be regarded as a rare event in reality.

6.2.2 Adversary Model

In our adversary model, we consider a strong adversary $\mathscr{A}$, whose goal is to identify privacy of as many as possible users in the residential area. Here, "strong" means that the adversary $\mathscr{A}$ is not only able to eavesdrop the communication flows but also capable of launching the following attacks: (a) $\mathscr{A}$ can directly compromise a user for privacy disclosure. However, since the number of users is large, $\mathscr{A}$ is discouraged to use this method to get more users' privacy; (b) $\mathscr{A}$ can deploy some undetectable malwares to the GW for privacy disclosure; and (c) $\mathscr{A}$ may compromise no more than d servers for privacy disclosure, where $d = \lceil k/2 \rceil - 1$. Of course, the adversary $\mathscr{A}$ can also launch some active attacks for goals other than privacy disclosure. However, since we primarily focus on privacy in this work, those attacks are beyond the scope of this chapter.

6.2.3 Security Requirements

The goal of the adversary $\mathscr{A}$ aims to break privacy of as many as possible users. To counteract the adversary's goal, our security requirements are as follows:

- Even though $\mathscr{A}$ can eavesdrop the communication flows, it cannot disclose users' private usage data.

- Even though $\mathscr{A}$ can compromise some users, it still cannot disclose other users' private usage data.
- Even though $\mathscr{A}$ can deploy some undetectable malwares to the GW, it still cannot disclose users' private usage data.
- Even though $\mathscr{A}$ can compromise d servers, it still cannot disclose users' private usage data.

6.2.4 Design Goal

Under the aforementioned system model, adversary model and security requirements, our design goal is to develop an efficient privacy-preserving aggregation scheme with fault tolerance for smart grid communications. Specifically, the following three objectives should be achieved.

- *The security requirements should be satisfied in the proposed aggregation scheme*. As stated above, if the smart grid does not consider the security, the residential users' privacy could be disclosed, and the smart grid cannot step into its flourish. Therefore, the proposed scheme should achieve the above security requirements accordingly.
- *The communication-effectiveness should be achieved in the proposed aggregation scheme*. Although the communication between the GW and the CC is featured with high-bandwidth and low-latency, to support a large number of residential users' reports to the CC at almost the same time, the proposed aggregation scheme should also consider the communication-effectiveness, so that the near real-time user reports can be efficiently transmitted to the CC.
- *The fault tolerance should be guaranteed in the proposed aggregation scheme*. Since some smart meters could be malfunctioning and d servers could possibly be compromised by the adversary $\mathscr{A}$, the proposed aggregation scheme should also be fault-tolerant, i.e., $k - d$ uncompromised servers can still recover the aggregated data from non-malfunctioning smart meters.

6.3 Proposed PDAFT Scheme

In this section, we propose our privacy-preserving data aggregation scheme with fault tolerance, called PDAFT, for smart grid communications. The proposed PDAFT scheme mainly consists of the following five phases: system initialization, user report generation, privacy-preserving report aggregation, secure report reading and fault tolerance handling.

6.3.1 System Initialization

We assume that the single trusted authority (TA) can bootstrap the whole system. Specifically, in the system initialization phase, given the security parameters κ and $d = \lceil k/2 \rceil - 1$, where k is the number of servers at the CC, TA first calculates and publishes the Paillier PKE's public key $(N = pq, g)$, where p, q are two large primes with $|p| = |q| = \kappa$, and $p = 2p' + 1, q = 2q' + 1$ where p' and q' are also large primes. Let $M = p'q'$. And then TA computes Paillier private key $SK = (\lambda, \alpha)$ with $\lambda = \text{lcm}(p-1, q-1)$ and $\alpha = (\Delta \cdot L(g^{\lambda} \mod N^2))^{-1} \mod N$, where $\Delta = k!$. The TA also defines a public cryptographic hash function $H : \{0,1\}^* \rightarrow \mathbb{Z}^*_{N^2}$ and publishes it. Finally, the TA needs to assign key materials to users $\mathbb{U} = \{U_1, U_2, \cdots, U_n\}$ and servers $\mathbb{S} = \{S_1, S_2, \cdots, S_k\}$ by the following steps.

Step 1: For each user $U_i \in \mathbb{U}$, TA first chooses a random number $k_{ui} \in \mathbb{Z}_N$ and assigns k_{ui} as U_i's private key.

Step 2: TA computes $\theta \in \mathbb{Z}_N$ such that

$$\theta + \sum_{i=1}^{n} k_{ui} = 0 \mod N \tag{6.1}$$

and randomly generates two secret polynomial functions of degree d as follows,

$$F(x) = \lambda + a_1 x + \cdots + a_d x^d \mod 4MN, \tag{6.2}$$

$$G(y) = \lambda \cdot \theta + b_1 y + \cdots + b_d y^d \mod 4MN, \tag{6.3}$$

where $a_i, b_i \in \mathbb{Z}_{MN}$ for $i = 1, 2, \cdots, d$.

Step 3: For each server $S_j \in \mathbb{S}$, TA first calculates the values of $F(j), G(j)$, and assigns

$$\{F(j), G(j), \alpha\} \tag{6.4}$$

as S_j's private key.

6.3.2 User Report Generation

Assume the reporting time points of users are defined as $\mathbb{T} = \{t_1, t_2, \ldots, t_{max}\}$ for some period. In order to report the nearly real-time residential users' usage data every 15 min, each user $U_i \in \mathbb{U}$ collects its usage data $m_{i,\gamma} \in \mathbb{Z}_N$ at time point $t_\gamma \in \mathbb{T}$ simultaneously, and performs the following steps:

Step-1: U_i first computes the hash value $h_\gamma = H(t_\gamma)$, and then calculates the ciphertext

$$C_{i,\gamma} = g^{m_{i,\gamma}} \cdot h_\gamma^{k_{ui}} \mod N^2. \tag{6.5}$$

Step-2: Finally, U_i reports $C_{i,\gamma}$ to the GW via WiFi.

6.3.3 Privacy-Preserving Report Aggregation

After receiving total n encrypted usage data $C_{i,\gamma}$ for $i = 1, 2, \ldots, n$, the GW performs the following steps for privacy-preserving report aggregation:

Step-1: The GW first computes the aggregated and encrypted data C_γ on $C_{1,\gamma}, C_{2,\gamma}, \ldots, C_{n,\gamma}$ as

$$C_\gamma = \prod_{i=1}^{n} C_{i,\gamma} \cdot R_\gamma^N = g^{\sum_{i=1}^{n} m_{i,\gamma}} \cdot h_\gamma^{\sum_{i=1}^{n} k_{ui}} \cdot R_\gamma^N \mod N^2, \tag{6.6}$$

where $R_\gamma \in \mathbb{Z}_N^*$ is a random number.

Step-2: Then, the GW reports the aggregated and encrypted data C_γ at time point t_γ to the CC.

Note that, if the smart meters of some users $\widehat{\mathbb{U}} \subset \mathbb{U}$ do not work, i.e., $\widehat{\mathbb{U}}$ won't report their data at time point t_γ. The GW aggregates the reported data

$$\widehat{C}_\gamma = \prod_{U_i \in \mathbb{U}/\widehat{\mathbb{U}}} C_{i,\gamma} \cdot R_\gamma^N = g^{\sum_{U_i \in \mathbb{U}/\widehat{\mathbb{U}}} m_{i,\gamma}} \cdot h_\gamma^{\sum_{U_i \in \mathbb{U}/\widehat{\mathbb{U}}} k_{ui}} \cdot R_\gamma^N \mod N^2, \tag{6.7}$$

then sends $\widehat{C}_\gamma$ and $\widehat{\mathbb{U}}$ to the CC.

6.3.4 Secure Report Reading

Upon receiving C_γ at time point t_γ at CC, $d + 1$ working servers $\mathscr{S} \subset \mathbb{S}$ are randomly chosen to decrypt the aggregated data. Specifically, each server $S_j \in \mathscr{S}$ first computes

$$\beta_j = \Delta \prod_{i \in \mathscr{S}, i \neq j} \frac{i}{i - j} \in \mathbb{Z}, \tag{6.8}$$

where $\Delta = k!$, k is the number of servers, then generates

$$D_{j,\gamma} = C_\gamma^{\beta_j F(j)} \cdot H(t_\gamma)^{\beta_j G(j)} \mod N^2. \tag{6.9}$$

After that, one of the servers $\mathscr{S}$ collects all the $D_{j,\gamma}$ for each $S_j \in \mathscr{S}$ and computes

$$\mathbf{C}_\gamma = \prod_{S_j \in \mathscr{S}} D_{j,\gamma} \mod N^2, \tag{6.10}$$

where $\mathbf{C}_\gamma$ is implicitly formed by

$$
\begin{aligned}
\mathbf{C}_\gamma &= \prod_{S_j \in \mathscr{S}} D_{j,\gamma} \mod N^2 \\
&= \prod_{S_j \in \mathscr{S}} C_\gamma^{\beta_j F(j)} \cdot H(t_\gamma)^{\beta_j G(j)} \mod N^2 \\
&= \prod_{S_j \in \mathscr{S}} C_\gamma^{\beta_j F(j)} \cdot \prod_{S_j \in \mathscr{S}} h_\gamma^{\beta_j G(j)} \mod N^2 \\
&= C_\gamma^{\sum_{S_j \in \mathscr{S}} \beta_j F(j)} \cdot h_\gamma^{\sum_{S_j \in \mathscr{S}} \beta_j G(j)} \mod N^2 \\
&= C_\gamma^{\Delta\lambda} \cdot h_\gamma^{\Delta\lambda\theta} = (C_\gamma \cdot h_\gamma^{\theta})^{\Delta\lambda} \mod N^2 \\
&= \left(g^{\sum_{i=1}^n m_{i,\gamma}} \cdot h_\gamma^{\sum_{i=1}^n k_{ui}+\theta} \cdot R_\gamma^N\right)^{\Delta\lambda} \mod N^2 \\
&\xrightarrow{\sum_{i=1}^n k_{ui}+\theta=0 \bmod N \Rightarrow \sum_{i=1}^n k_{ui}+\theta=\mu N \text{ for some } \mu} \\
&= \left(g^{\sum_{i=1}^n m_{i,\gamma}} \cdot (h_\gamma^{\mu} \cdot R_\gamma)^N\right)^{\Delta\lambda} \mod N^2 \\
&= g^{\Delta\lambda \sum_{i=1}^n m_{i,\gamma}} \cdot (h_\gamma^{\Delta\mu} \cdot R_\gamma^{\Delta})^{\lambda N} \mod N^2 \\
&= g^{\Delta\lambda \sum_{i=1}^n m_{i,\gamma}} \mod N^2.
\end{aligned}
\tag{6.11}
$$

Finally, the aggregated value $\sum_{i=1}^n m_{i,\gamma}$ can be obviously calculated by Paillier decryption, i.e.,

$$
\sum_{i=1}^n m_{i,\gamma} = L(\mathbf{C}_\gamma) \cdot \alpha \mod N. \tag{6.12}
$$

6.3.5 Fault Tolerance Handling

As discussed in our system model, no more than d servers could be compromised. However, since $d = \lceil k/2 \rceil - 1$, we still have $k - d \geq d + 1$ working servers to recover the aggregated data $\sum_{i=1}^n m_{i,\gamma}$.

If the smart meters of some users $\widehat{\mathbb{U}} \subset \mathbb{U}$ do not work, $\widehat{C}_\gamma$ and $\widehat{\mathbb{U}}$ will be received at the CC. Then, the TA is required to provide a dummy ciphertext related to users $\widehat{\mathbb{U}}$ at time point t_γ. Concretely, as the TA knows the private keys of users $\widehat{\mathbb{U}}$, the TA can calculate

$$
\overline{C}_\gamma = \prod_{U_i \in \widehat{\mathbb{U}}} H(t_\gamma)^{k_{ui}} = H(t_\gamma)^{\sum_{U_i \in \widehat{\mathbb{U}}} k_{ui}} \mod N^2. \tag{6.13}
$$

Then, by combining $\overline{C}_\gamma$ and $\widehat{C}_\gamma$, we have

$$C_\gamma = \widehat{C}_\gamma \cdot \overline{C}_\gamma = g^{\sum_{U_i \in \mathbb{U}/\widehat{\mathbb{U}}} m_{i,\gamma}} \cdot h_\gamma^{\sum_{U_i \in \mathbb{U}} k_{ui}} \bmod N^2. \tag{6.14}$$

Similar as the procedures in *Secure Report Reading* part, $d + 1$ working servers can finally recover the aggregated data $\sum_{U_i \in \mathbb{U}/\widehat{\mathbb{U}}} m_{i,\gamma}$.

6.4 Security Analysis

In this section, we will discuss the security issues involved in the proposed scheme, specifically, to protect the users' electricity usage privacy against a strong adversary $\mathcal{A}$.

- *The users' electricity usage privacy is protected from eavesdropping.* As stated in our security model, an adversary $\mathcal{A}$ may reside in the residential area to eavesdrop the communication flows from users to the GW. Suppose the $\mathcal{A}$ has eavesdropped a ciphertext of user U_i at time point t_γ, i.e. $g^{m_{i,\gamma}} \cdot h_\gamma^{k_{ui}}$. Since the electricity usage $m_{i,\gamma}$ within 15 min is probably a small value, the adversary $\mathcal{A}$ may try to launch a brute-force attack by exhaustedly testing each possible value of $m_{i,\gamma}$, but before that, $\mathcal{A}$ needs to know $h_\gamma^{k_{ui}}$ first, which is impossible if k_{ui} is unknown to $\mathcal{A}$. Thus the privacy of users' electricity usage is guaranteed.
- *The uncompromised users' electricity usage will not be revealed.* In our system, we consider that the adversary $\mathcal{A}$ can compromise some of the users, in that case, the privacy of compromised users is fully exposed. However, since there are a large number of users, the adversary $\mathcal{A}$ is discouraged to use this inefficient method to disclose more users' privacy. Instead the adversary $\mathcal{A}$ may try to threaten the uncompromised users' privacy by utilizing the secret information he obtained from the compromised ones, i.e. their private keys. Nevertheless, this attack won't success either since the privacy key of each user is randomly chosen by the TA, knowing one user's private key reveals nothing about another one's. Moreover, even if the adversary $\mathcal{A}$ compromises $n - 1$ users and obtains their private keys, he still can not reveal the last user's private key and electricity usage. Thus the privacy of uncompromised users is also preserved.
- *The users' private usage and sum usage data will not be disclosed at the GW.* At each time point t_γ, the GW collects all users' ciphertexts and directly aggregates them into one single ciphertext. Thus if the adversary $\mathcal{A}$ deployed some undetectable malwares into the GW, he could only get the ciphertexts of all users and the aggregated one. Since the GW does not decrypt any user's electricity usage data, the $\mathcal{A}$ still cannot get any user's private usage data. Moreover, the aggregated ciphertext is $C_\gamma = g^{\sum_{i=1}^n m_{i,\gamma}} \cdot h_\gamma^{\sum_{i=1}^n k_{ui}} \bmod N^2$, which has the same form of one user's ciphertext $g^{m_{i,\gamma}} \cdot h_\gamma^{k_{ui}}$. Since the sum of all users' private key is not known to the adversary $\mathcal{A}$, similarly, the adversary $\mathcal{A}$ can

neither disclose the sum usage of all users. Thus the users' electricity usage data privacy can be ensured even though the GW is deployed some undetectable malwares by the adversary.

- *The adversary cannot obtain users' private usage and sum usage data even if d servers are compromised.* For the system that has only one server, if the server occurs malfunction or it is compromised by the adversary, the whole system will suffer from the single point of failure. A simple countermeasure for this problem is to deploy a replica server, if one of them occurs malfunction, the other can keep the whole system working on. However, if a strong adversary compromises one of the two servers, he can then obtain all the secret information of the system. Varying from the simple method, in our system, $k \geq 3$ servers are deployed to work synergistically, and are assigned different private keys $\{F(j), G(j), \alpha\}, j = 1, \ldots, k$. In this case, even if no more than $d = \lceil k/2 \rceil - 1$ servers have occurred malfunction or have been compromised, the system can still work properly and keep the secret information from disclosing to the adversary. Specifically, suppose the strong adversary $\mathscr{A}$ compromises d servers, w.l.o.g $\{U_1, U_2, \cdots, U_d\}$, and obtains their private keys $\{F(j), G(j), \alpha\}, j = 1, \ldots, d$, where $F(j)$ is the share of λ and $G(j)$ is the share of $\lambda\theta$. Based on the "all or nothing" property of secret sharing [14], at least $d + 1$ shares are necessary for recovering λ or $\lambda\theta$, thus the adversary $\mathscr{A}$ will get no information about the secret λ or $\lambda\theta$. Similarly, to successfully decrypt the sum usage of users, $d + 1$ servers are needed to create $d + 1$ decryption shares $D_{j,\gamma} = C_\gamma^{\beta_j F(j)} \cdot H(t_\gamma)^{\beta_j G(j)} \bmod N^2$. Although the adversary can also create d decryption shares with keys from the compromised servers, d shares are not enough to recover C_γ^λ or $H(t_\gamma)^{\lambda\theta}$, thus the adversary $\mathscr{A}$ still can not decrypt the aggregated ciphertext to acquire the sum usage of users. On the other hand, since our system is fault-tolerant, only if the compromised or malfunctioned servers do not exceed d ones, we still have $k-d \geq d+1$ working servers to keep the system processing normally. In addition, our system is also fault-tolerant of users' malfunction. Specifically, if a user U_i occurs malfunction, since the TA knows all users' private key, TA can assist the servers decrypting the aggregated ciphertext by providing a dummy ciphertext $h_\gamma^{k_{ui}}$. Even if the adversary captures the dummy usage data, the private key k_{ui} of user U_i will not expose to the adversary $\mathscr{A}$ since the discrete logarithm problem is computationally difficult for the adversary $\mathscr{A}$. Moreover, since knowing d server keys are useless for the adversary $\mathscr{A}$ to recover any user's private key, as aforementioned, the adversary $\mathscr{A}$ can neither reveal any user's private usage, even if $\mathscr{A}$ has compromised d servers. Based on the above discussion, the users' private usage and sum usage data are both protected from the strong adversary.

6.5 Performance Evaluation

In this section, we evaluate the performance of the proposed PDAFT scheme in terms of the communication overhead in the smart grid communication. We compare our proposed scheme with the one proposed by Erkin and Tsudik [12] which also supports privacy-preserving aggregation, i.e. the individual usage will not be revealed by the adversary. Erkin and Tsudik's scheme not only supports *spacial aggregation*, i.e. the aggregation of different users' data at the same time point, but also supports *temporal aggregation*, i.e. the aggregation of the same user's data at different time points. In addition, since their scheme is distributed, it can easily support dynamic users, i.e. user addition or user removal. Actually, our scheme can also support temporal aggregation and dynamic users, we first give the corresponding extensions below, then compare the communication overhead of the two schemes.

6.5.1 Extension to Support Temporal Aggregation

In the smart grid communication, the aggregation of different users' data at the same time point can be named as "spacial aggregation" [12], it is widely used in leakage detection, fraud detection, forecasting, etc. However, sometimes we also need to calculate another kind of aggregation called "temporal aggregation", e.g. in pricing and billing schemes [15], which aggregates one user's data of a number of continuous time points. Evidently, in the temporal aggregation we also need to protect individual user's usage privacy. Our proposed protocol can be easily modified to support privacy-preserving temporal aggregation.

Assume a temporal aggregation cycle contains b time points, i.e. from t_1 to t_b, from t_{b+1} to t_{2b}, and so on. Note that the aggregation cycle b should not be too small or the user behavior privacy may also be violated. We take the first cycle as an exam to present our extension as follows. For user U_i, as in our proposed scheme, its usage is encrypted as $g^{m_{i,\gamma}} \cdot h_{\gamma}^{k_{ui}}$ for time point t_γ. We keep the ciphertexts consistent for $\gamma = 1, 2, \ldots, b-1$, but modify the last one at time point t_b as $\widetilde{C}_{i,b} = g^{m_{i,b}} \cdot \tilde{h}_b^{k_{ui}} \cdot r_i^N$, where $r_i \in \mathbb{Z}_N^*$ is a random number and

$$\tilde{h}_b = \left(\prod_{\gamma=1}^{b-1} h_\gamma \right)^{-1} \mod N^2.$$

For the temporal aggregation, the GW collects b ciphertexts and aggregates them as follows.

$$
\begin{aligned}
\widetilde{C}_i &= \widetilde{C}_{i,b} \cdot \prod_{\gamma=1}^{b-1} C_{i,\gamma} \bmod N^2 \\
&= g^{\sum_{\gamma=1}^{b} m_{i,\gamma}} \cdot \left(\prod_{\gamma=1}^{b-1} h_\gamma \right)^{k_{ui}} \cdot \left(\tilde{h}_b\right)^{k_{ui}} \cdot r_i^N \bmod N^2 \\
&= g^{\sum_{\gamma=1}^{b} m_{i,\gamma}} \cdot r_i^N \bmod N^2.
\end{aligned}
\tag{6.15}
$$

Then the GW reports the aggregated ciphertext $\widetilde{C}_i$ to the servers at CC. To decrypt the sum usage for U_i, $d+1$ working servers $\mathscr{S}$ compute their decryption shares $\widetilde{D}_{j,i} = \widetilde{C}_i^{\beta_j F(j)}$ respectively, and aggregate them together to get

$$
\begin{aligned}
\widetilde{\mathbf{C}}_i &= \prod_{S_j \in \mathscr{S}} \widetilde{D}_{j,i} \bmod N^2 \\
&= \widetilde{C}_i^{\sum_{S_j \in \mathscr{S}} \beta_j F(j)} \bmod N^2 \\
&= \left(g^{\sum_{\gamma=1}^{b} m_{i,\gamma}} \cdot r_i^N \right)^{\lambda} \bmod N^2.
\end{aligned}
\tag{6.16}
$$

Finally, the aggregated value $\sum_{\gamma=1}^{b} m_{i,\gamma}$ can be obviously calculated by Paillier decryption, i.e.,

$$
\sum_{\gamma=1}^{b} m_{i,\gamma} = L(\widetilde{\mathbf{C}}_i) \cdot \alpha \bmod N. \tag{6.17}
$$

Note that since we modified the ciphertexts of users at time point t_b, we still need to check whether the spatial aggregation can be processed correctly. The GW aggregates the ciphertexts as $C_b = \prod_{i=1}^{n} \widetilde{C}_{i,b} = g^{\sum_{i=1}^{n} m_{i,b}} \cdot \tilde{h}_b^{\sum_{i=1}^{n} k_{ui}} \cdot \left(\prod_{i=1}^{n} r_i\right)^N \bmod N^2$. Comparing C_b with C_γ in our original scheme, there is an addition factor $\left(\prod_{i=1}^{n} r_i\right)^N$, but it will not influence the decryption result. Since similar as Eq. (6.11), the final ciphertext is $\mathbf{C}_b = \left(g^{\sum_{i=1}^{n} m_{i,b}} \cdot (\tilde{h}_b^{\mu} \cdot \prod_{i=1}^{n} r_i)^N \right)^{\lambda} \bmod N^2$, which can also be correctly decrypted to get the sum usage $\sum_{i=1}^{n} m_{i,b}$. Thus our extended scheme can support spatial and temporal privacy-preserving aggregation simultaneously like Erkin and Tsudik's scheme does.

6.5.2 Extension to Support Dynamic Users

Users in the system may not always remain unchangeable, i.e. the user set $\mathbb{U}$ may sometimes add a new user or revoke an old one. Erkin and Tsudik's scheme is distributed and all users will exchange random numbers at each time point, thus it is adaptive for user addition and removal. Our scheme can also be modified to have the same property. Since only the TA knows each user's private key, when a set of new users $\mathcal{U}_a$ are added into the system and a set of old users $\mathcal{U}_b$ are removed, the TA generates each $U_a \in \mathcal{U}_a$ a new private key k_{ua} and computes

$$\theta' = \theta - \sum_{U_a \in \mathcal{U}_a} k_{ua} + \sum_{U_b \in \mathcal{U}_b} k_{ub}. \tag{6.18}$$

Then the TA updates the secret polynomial $G(y)$ as

$$G'(y) = \lambda \cdot \theta' + b'_1 y + \cdots + b'_d y^d, \tag{6.19}$$

where $b'_i \in \mathbb{Z}_{N^2}$ for $i = 1, 2, \cdots, d$, and updates $G(j)$ in each server S_j's private key as $G'(j)$. In this case, the servers are able to decrypt the aggregated usage of new user set $\mathbb{S} + \mathcal{U}_a - \mathcal{U}_b$. The correctness can be checked by substituting $G'(j)$ for $G(j)$ in Eq. (6.11) in the *Secure Report Reading* part.

Although our extension needs the servers to update their privacy keys when the user set is changed, we believe the cost is acceptable. Since we needn't to update users' private keys and the number of servers k is much less than that of users n. In addition, user set changing in the smart grid may not happen frequently in reality.

6.5.3 Communication Overhead Comparison

The communication overhead of the proposed PDAFT scheme can be considered in two metrics, the individual user communication and the overall communication. We first consider the communication of individual users, where each user generates his data report and delivers it to the local GW. The size of user U_i's report is $|C_{i,\gamma}| = 2048$ bits if we choose the Paillier parameter $\kappa = 512$. Note that we ignore other parts of the packet such as user ID and timestamp which are relatively short compared to the report. Let $S_N = 1024$ bits, since each user in PDAFT only needs to send one report at a time point, the user communication overhead is $S_U = 2 \cdot S_N$. By contrast, in Erkin and Tsudik's scheme users need to exchange random numbers and broadcast reports to all other users at each time point. Specifically, each user will send/receive $(n-1)$ random numbers of size S_N and broadcast/collect $(n-1)$ reports of size $2 \cdot S_N$. Thus the total communication overhead of each user is $S'_U = 6(n-1) \cdot S_N$. We plot the user communication overhead of both schemes in terms of user number n and running time γ, as shown in Fig. 6.3. It can be seen that the proposed PDAFT scheme achieves much lower user communication overhead compared to Erkin and Tsudik's scheme.

Next, we consider the overall communication of both PDAFT and Erkin and Tsudik's scheme. In PDAFT, the GW collects n users' reports and aggregates them into a single one C_γ whose size is just $2 \cdot S_N$, then the GW sends C_γ to the servers at CC. Summing up all users' communication overhead and that from GW to CC, the

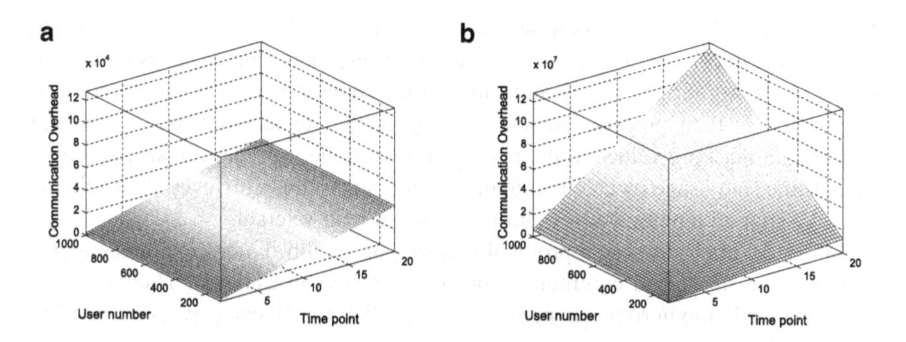

Fig. 6.3 Communication overhead of individual users. (**a**) Individual communication overhead of PDAFT. (**b**) Individual communication overhead of Erkin and Tsudik's scheme

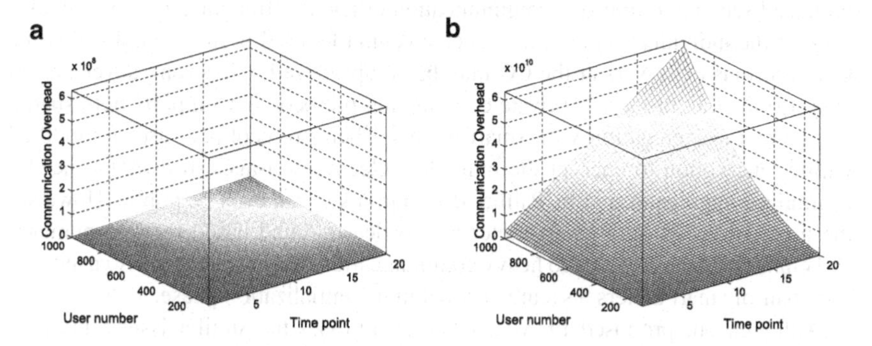

Fig. 6.4 Overall communication overhead of (**a**) PDAFT and (**b**) Erkin and Tsudik's scheme

overall communication overhead of PDAFT at one time point is $S_O = (2n + 2) \cdot S_N$. On the other hand, in Erkin and Tsudik's scheme the users do not send reports to the GW but broadcast them to all users, thus the overall communication overhead at each time point is $S'_O = 3n(n-1) \cdot S_N$. In Fig. 6.4, we further plot the overall communication overhead in terms of user number n and running time γ. It is shown that the PDAFT scheme significantly reduces the overall communication overhead in the smart grid communication.

Timeliness is essential for smart grid communications. From the above analysis, the proposed PDAFT scheme is indeed efficient in terms of communication cost, which can also significantly reduce the transmitting time for data collection, especially when the user set is large.

6.6 Related Work

In smart grid communication, aggregation techniques can obviously decrease computation and communication overhead of the CC, assist the CC to provide effective and low-latency services such as leakage detection and forecasting. Therefore, in this section, we briefly discuss some other research works [8, 9, 12] closely related to our scheme. In [9], Shi et al. propose an aggregation scheme that protects individual user's electricity usage privacy and supports differential privacy with an untrusted aggregator. In their proposed scheme, the aggregator collects all users' data and cancels the masking values in the aggregation with the aggregator's key, and then obtains the sum usage by calculating the discrete logarithm. However, their scheme does not support temporal aggregation, and is not fault-tolerant.

In [8], Lu et al. propose a protocol that supports multi-dimensional data aggregation in the smart grid communication (i.e., PPMDA in Chap. 3). Their protocol also uses the homomorphic property of Paillier PKE to achieve privacy-preserving aggregation against the semi-honest aggregator. The protocol significantly reduces the computation and communication overhead of the system by encrypting multi-dimensional data into one single ciphertext, which can also be applied in our proposed scheme to improve communication efficiency. But since the CC can also decrypt the individual reports, the scheme cannot be used in the scenarios that the CC is curious or servers at the CC may be compromised by a strong adversary. In [12], Erkin et al. propose a private computation protocol that supports both spatial and temporal aggregation with smart meters. Their protocol also uses a modified Paillier encryption to encrypt the individual electricity usages. It does not need a centralized aggregator and its Paillier decryption key can even be public. However, the protocol needs each smart meter to broadcasts round-based random numbers and ciphertexts which leads to heavy communication overhead, and it also needs for each pair of smart meters a secure channel in the initialization phase.

Although our proposed PDAFT scheme addresses the similar issues, i.e. providing communication-efficient and privacy-preserving aggregation, and supporting both spatial and temporal summation, in contrast to the above works, our research focuses still have some differences: (a) we propose our aggregation protocol in a more challenging threaten model that the adversary could compromise some of the servers at the CC and obtain the private keys, and we not only protects the individual electricity usage from disclosing to the adversary, but also prevent the adversary to know the sum usage of the RA; and (b) we take malfunction of both users and servers into consideration that additionally improves the reliability and practicability of our aggregation scheme by supporting fault tolerance.

6.7 Summary

In this chapter, we have proposed a privacy-preserving aggregation protocol that supports both spatial and temporal aggregation of user electricity usages. We protect the user privacy under a strong adversary model in which the adversary can compromise some of the servers at the CC to obtain their private keys. The capability of dealing with malfunction of both users and servers makes our aggregation system more reliable and practical in smart grid communication. Detailed security analysis shows that the proposed scheme can prevent a strong global adversary from learning either the individual user data or the sum data of the RA. In addition, through extensive performance evaluation, we have also demonstrated that the proposed scheme can save much communication overhead of residential users in the smart grid.

References

1. L. Chen, R. Lu, and Z. Cao, "PDAFT: A privacy-preserving data aggregation scheme with fault tolerance for smart grid communications," *Peer-to-Peer Networking and Applications*, vol. 8, no. 6, pp. 1122–1132, 2015. [Online]. Available: http://dx.doi.org/10.1007/s12083-014-0255-5
2. H. Liang, B. J. Choi, A. Abdrabou, W. Zhuang, and X. S. Shen, "Decentralized economic dispatch in microgrids via heterogeneous wireless networks," *IEEE Journal on Selected Areas in Communications*, vol. 30, no. 6, pp. 1061–1074, 2012.
3. K. Kursawe, G. Danezis, and M. Kohlweiss, "Privacy-friendly aggregation for the smart-grid," in *PETS*, 2011, pp. 175–191.
4. C. Efthymiou and G. Kalogridis, "Smart grid privacy via anonymization of smart metering data," in *SmartGridComm*, 2010, pp. 238–243.
5. K. Alharbi and X. Lin, "Lpda: A lightweight privacy-preserving data aggregation scheme for smart grid," in *WCSP*, 2012, pp. 1–6.
6. W. Jia, H. Zhu, Z. Cao, X. Dong, and C. Xiao, "Human-factor-aware privacy preserving aggregation in smart grid," *IEEE System Journal*, to appear.
7. F. Li, B. Luo, and P. Liu, "Secure information aggregation for smart grids using homomorphic encryption," in *SmartGridComm*, 2010, pp. 327–332.
8. R. Lu, X. Liang, X. Li, X. Lin, and X. Shen, "Eppa: An efficient and privacy-preserving aggregation scheme for secure smart grid communications," *IEEE Trans. Parallel Distrib. Syst.*, vol. 23, no. 9, pp. 1621–1631, 2012.
9. E. Shi, T.-H. H. Chan, E. G. Rieffel, R. Chow, and D. Song, "Privacy-preserving aggregation of time-series data," in *NDSS*, 2011.
10. V. Rastogi and S. Nath, "Differentially private aggregation of distributed time-series with transformation and encryption," in *SIGMOD*, 2010, pp. 735–746.
11. F. D. Garcia and B. Jacobs, "Privacy-friendly energy-metering via homomorphic encryption," in *STM*, 2010, pp. 226–238.
12. Z. Erkin and G. Tsudik, "Private computation of spatial and temporal power consumption with smart meters," in *ACNS*, 2012, pp. 561–577.
13. "Wireless medium access control (mac) and physical layer (phy) specifications for low-rate wireless personal area networks (wpans) amendment 4: Physical layer specifications for low data rate wireless smart metering utility networks," *IEEE Std. P802.15.4g/D4 part 15.4*, Apr. 2011.
14. A. Shamir, "How to share a secret," *Commun. ACM*, vol. 22, no. 11, pp. 612–613, 1979.
15. X. Liang, X. Li, R. Lu, X. Lin, and X. Shen, "Udp: Usage-based dynamic pricing with privacy preservation for smart grid," *IEEE Trans. Smart Grid*, vol. 4, no. 1, pp. 141–150, 2013.

Chapter 7
Differentially Private Data Aggregation with Fault Tolerance

In this chapter, we introduce a new secure data aggregation scheme, named DPAFT, for secure smart grid communications [1]. In specific, the proposed DPAFT can not only support fault tolerance but also resist against the differential attacks in smart grid communications.

7.1 Introduction

Due to the lack of effective realtime diagnosis and efficient load imbalances, the conventional power grid is not always so stable, and would be disastrous to our daily lives. For example, in August 2003, more than 100 power plants were penetrated and tens of millions of people were paralyzed by the North America electrical blackout [2]. Facing such challenges, smart grid, a new promising generation of power system, emerged and has been paid great attention not only by government but also by academia and industry [3–6].

Compared with traditional power grid, smart grid has combined many progressive technologies, *e.g.*, data sensing and control, information collection and monitoring, into the traditional power grid, enabling the power distribution to be more efficient and reliable from power generation, transmission, and distribution to customers consumption, and supports the renewable energy [7–9]. By deploying various sensors along the two-way flows, *i.e.*, the electricity flow and the communication flow, a huge amount of real-time information is reported and collected to the control center (CC) for timely monitoring and analyzing the health of the power grid, as shown in Fig. 7.1. Specifically, all the electric appliances in the residential user's home are connected to a central element, smart meter, which periodically collects and reports the power consumption of appliances to the local area gateway (GW). Then the GW aggregates and forwards the data to the CC for further analysis

R. Lu, *Privacy-Enhancing Aggregation Techniques for Smart Grid Communications*, Wireless Networks, DOI 10.1007/978-3-319-32899-7_7

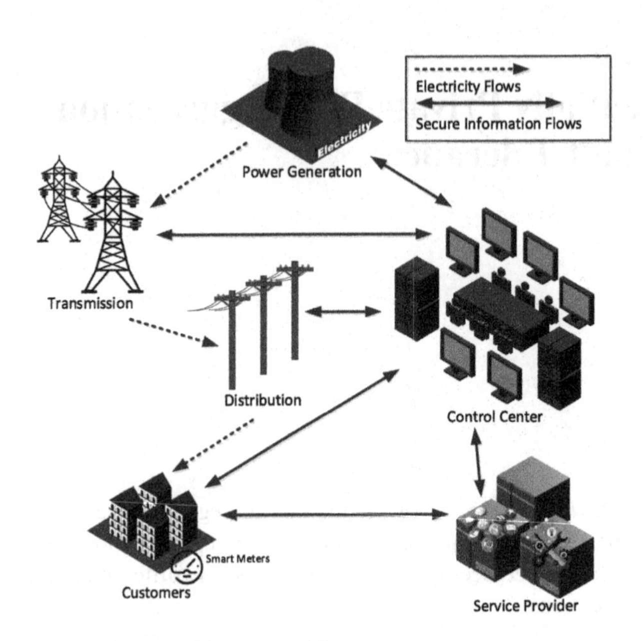

Fig. 7.1 The conceptual smart grid system architecture

and processing, *e.g.*, making real-time power pricing decisions [10], detecting power fraud or leakage [11], *etc.*

However, the real-time usage data, *e.g.*, collected every 15 min, contain personalized power usage patterns which are highly relevant to users' privacy, thus they must be protected from unauthorized entities. Up to now, many data aggregation schemes [2, 12–16] have been proposed to preserve individual user privacy in smart grid.

Most of them [2, 12, 14, 15] use the homomorphic encryption techniques [17] to encrypt user's data so that the *semi-trust* aggregator (*e.g.,* the GW) can aggregate all users' data without decryption. However, they only consider the protection of users' privacy against the GW (aggregator), while the CC, if considered under the *honest-but-curious* model, is still able to learn individual user's data, as the keys owned by the CC may be not only utilized to decrypt the aggregated data, but also used to reveal any user's electricity usage. In order to prevent the curious CC from disclosing individual user's privacy, other schemes [13, 18, 19] use a subtle key management technique, *i.e.*, the sum of all keys equals 0, to decentralize the power of the CC so as to enhance the security of the data aggregation protocols. However, one of the major limitations of these works is that they cannot support fault tolerance, *i.e.*, once one user fails to report, the whole data aggregation protocol is not workable. Therefore, fault tolerance is a big concern for smart grid communications, because smart meters, as low-cost devices and running in unprotected environments, are prone to failures.

Another challenging problem that each secure data aggregation scheme could face is the differential attack [20, 21]. The idea of differential attack is straight forward. Even if an aggregation scheme is secure, once the CC acquires the aggregations of n users and that of $n-1$ users, the privacy of the differential one will be leaked, although the aggregation of n users and that of $n-1$ users are both secure. Specifically, if an adversary acquires the aggregations of two similar data sets $D_U, D_V \subset U$ differing on at most one element, *i.e.*, $D_U = D_V + U_\alpha$, assume $A(D)$ denotes the sum aggregation of data set D, then $A(D_U) - A(D_V)$ is exactly the data of user U_α. This problem, the differential attack in smart grid, has been addressed in several literatures, such as [13, 14, 22–24]. However, most of them [13, 14, 22] are not fault tolerant. A handful of them [23, 24] can support fault tolerance, but either with low efficiency [23] or with unsatisfying utility (*i.e.*, with inappropriate error) [24], which makes them unpractical to real scenarios where the exact number of malfunctioning smart meters are unpredictable and sometimes could be large to some extent. Therefore, how to design an efficient, privacy-preserving, high utility (*i.e.*, low error), differentially private, and fault tolerant data aggregation for smart grid still deserves further investigations.

Motivated by the above-mentioned, in this chapter, we propose a novel *d*ifferentially *p*rivate data *a*ggregation with *f*ault *t*olerance (DPAFT) scheme, which is characterized by fault-tolerant for smart metering that can handle general measurements report failures while ensuring differential privacy with significantly improved efficiency and lower errors compared with the state of the art. Specifically, the main contributions of this chapter are three-fold.

- Firstly, inspired by the idea of Diffie-Hellman key exchange protocol, we propose a novel key management technique for handling fault tolerance in smart metering. Unlike those existing works [12–14, 23, 24], which depend on the restricted relation of $\sum_{i=0}^{n} s_i = 0$, an artful constraint relation $s_0 \cdot \sum_{i=1}^{n} s_i = 1$ is constructed, where s_0, and s_i, for $i = 1, \cdots, n$, are the private keys of the *CC*, and each residential user, respectively. With such novel constraint, the Diffie-Hellman keys like $Y_i = h^{s_0 s_i}$, for $i = 1, \cdots, n$, are issued to the CC, which are utilized to support fault tolerance of malfunctioning smart meters efficiently and flexibly.
- Secondly, observing the fact that users' private data may often suffer from the differential attacks, our enhanced DPAFT is designed to provide differential privacy by introducing distributed noise generation procedure. Compared with the state of the art differentially private smart grid aggregation protocol [24], our protocol is more efficient due to the elimination of heavy communication, computation, and storage overhead of *future-ciphertexts*, while still providing high utility (*i.e.*, low error) by adding Laplace noise via distributed manner.
- Finally, by improving the basic BGN PKE [25] to be more applicable to the practical scenarios, our DPAFT can be secure against much stronger adversary and highly efficient. Specifically, by hiding the private key p of the basic BGN PKE to the *CC* and introducing the blind factor t for the GW and the secret key r for the *CC*, respectively, the users' electricity usage privacy is protected in *honest-but-curious model*, *i.e.*, only all the participants conform the protocol and

collaboratively execute the procedures using their individual secret information, can the sum usage data be recovered. Any participant running in the *honest-but-curious* model cannot infer useful knowledge about any residential user's privacy.

The remainder of this chapter is organized as follows. In Sect. 7.2, we introduce our system model, attack model, and identify our design goal. Then in Sect. 7.3, we briefly review some preliminaries. After that, we present our basic and enhanced DPAFT in Sects. 7.4 and 7.5, respectively. Then, the security analysis is illustrated in Sect. 7.6 and the performance evaluations are given in Sect. 7.7. We also discuss the related work in Sect. 7.8. Finally, we draw our conclusions in Sect. 7.9.

7.2 Problem Formalization

In this section, we formalize system model, attack model, and identify our design goal.

7.2.1 System Model

Since residential users put particular emphases on their privacy when reporting their electricity consumptions to the control center (CC) in smart metering systems, and smart meters as inexpensive home devices, which are often deployed in unprotected environments and communicated over unreliable network channel, are unreliable and fail to report the usage data sometimes, in this work, we mainly put emphasis on how to report users' measurements to the CC in a privacy-preserving and fault-tolerant way. Specifically, in our system model, we consider a typical smart grid communication architecture for residential users, which includes a trusted authority (TA), a control center (CC), a residential gateway (GW), and a great number of residential users $U = \{U_1, U_2, \cdots, U_n\}$ in a residential area (RA), as shown in Fig. 7.2.

Trusted Authority (TA): The TA is a trustable and powerful entity in charge of management of the whole system. In general, after initializing the system, the TA will be off-line, *i.e.*, it won't directly participate in the users' data reporting unless some exceptions occur in reporting.

Control Center (CC): The CC is a highly-trusted entity, whose duty is to collect, process and analyze the nearly real-time data for providing reliable services for smart grid.

Gateway (GW): The GW is a powerful entity, which connects the CC and the residential users. The responsibility of the GW mainly includes the two aspects: aggregation and relaying. The responsibility of aggregation is to aggregate the

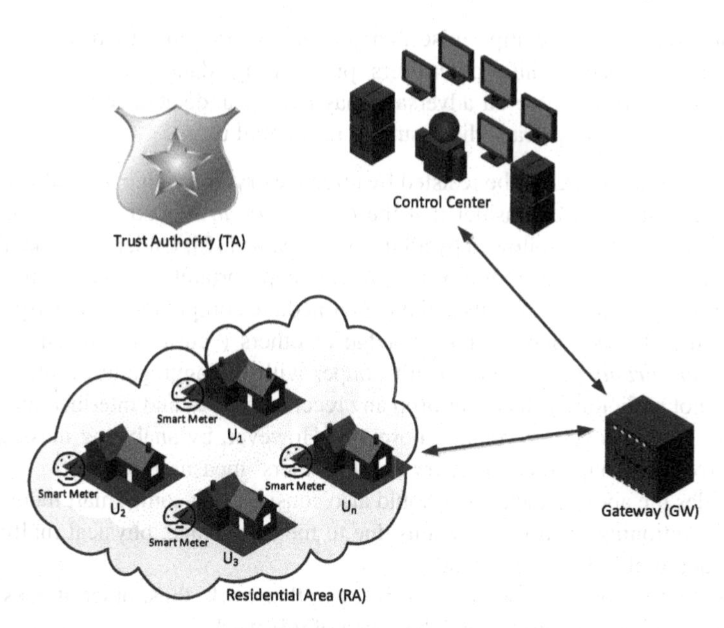

Fig. 7.2 System model under consideration

measurements from RA into an integrated one, while the responsibility of relaying is to assist in forwarding the communication flow between the CC and residential users in a secure way.

Residential Users $U = \{U_1, U_2, \cdots, U_n\}$: Each residential user $U_i \in U$ is equipped with a smart meter and various smart appliances to form a Home Area Network (HAN), which can collect the real-time measurements, and report them to the CC via the GW in a certain period, *e.g.*, every 15 min. Because smart meter is not as powerful as the GW, some meters could be malfunctioning occasionally, *i.e.*, they could stop reporting for a while and will be reset in a late time. Nevertheless, malfunction of smart meter can be regarded as a rare event in reality.

7.2.2 Attack Model

In our attack model, we consider the following three most frequently launched attacks in smart grid systems:

- *External attack,* where an adversary may compromise the privacy of residential users by eavesdropping the communication data from the residential users to the GW and those from the GW to the CC;
- *Internal attack*, where an adversary is usually the participants of the protocol including the GW or the CC, which could access or misuse the information of

residential users to compromise their privacy, or the curious residential users, who actively seek or infer other users' private usage data;
- *Malware attack*, where an adversary may deploy undetectable malwares to the GW or the CC for privacy disclosure of residential users.

The external attack can be resisted by effective cryptographic algorithms. As for the internal attack, we consider that the *honest-but-curious* adversary model, *i.e.*, semi-honest model, is followed by all the participants [15]. Under this model, all the participants are supposed to follow the protocol appropriately ("honest"); meantime, by keeping other parties' inputs and the intermediate computations, they try all sorts of measures to seek and infer knowledge of others ("curious"). In our scenario, *honest-but-curious* participants won't tamper with the aggregation protocols, *i.e.*, they do not maliciously distort or drop any received values and intermediate results, and they keep the system running normally. However, by analyzing messages and values routed through them, they try to infer others' measurements.

Besides the above attacks, we should also consider that some smart meters could be malfunctioning or in failure status due to meter wastage, physical malfunction, *etc.*, which will fail to report data.

Note that we primarily focus on privacy in this work, thus, other attacks except for privacy disclosure are beyond the scope of this work.

7.2.3 Design Goal

Considering the aforementioned system model and attack model, our design goal focuses on proposing an efficient differentially private data aggregation with fault tolerance for smart metering. Specifically, the following design goals are to be achieved:

- *Privacy-Preserving*: Firstly, an external attacker $\mathscr{A}$ cannot disclose users' private usage data even though $\mathscr{A}$ can eavesdrop the communication flows. Secondly, although $\mathscr{A}$ can deploy some undetectable malwares to the GW or the CC, it still cannot disclose users' private usage data. Thirdly, through eavesdropping and analyzing all the inputs, intermediate communication flows and outputs which are not of one's own, any participant running in the *honest-but-curious* adversary model cannot infer useful knowledge about residential users' privacy. Finally, $\mathscr{A}$ cannot launch differential attack to obtain the individual user's privacy successfully;
- *Fault Tolerance*: The system can still aggregate the data of functioning meters effectively and efficiently even in presence of malfunctioning ones;
- *Computation Efficiency:* The computation efficiency should be achieved in the proposed protocol to support thousands and millions of residential users' data aggregation.

7.3 Preliminary

In this section, we briefly recall *differential privacy* [20], which serves as the preliminaries for our differentially private data aggregation scheme with fault tolerance.

7.3.1 Differential Privacy

The ε-Differential Privacy (ε-DP). A randomized algorithm A is ε-differentially private if for any two data sets D_u and D_v differing on one element, and for all $R \subset Range(A)$, $Pr[A(D_u) \in R] \leq exp(\varepsilon) \cdot Pr[A(D_v) \in R]$.

Differential Privacy via Laplace Noise. It is proved by Dwork in [20] that adding i.i.d. Laplace noise Lap(λ) to the accurate result could achieve ε-differential privacy. The Laplace noise Lap(λ) is sampled from the Laplace distribution with the parameter λ defined as $\lambda = \frac{GS_f}{\varepsilon}$, where GS_f indicates the global sensitivity of the function f. In smart grid communications, f is the measurement of the user, so GS_f is the maximum amount that any user may consume during any reporting period. We usually call λ as the noise scale. A random variable following the Laplace distribution has expected absolute deviation λ and standard deviation $\sqrt{2}{\cdot}\lambda$. Consequently, the smaller ε is, the noisier the outputs, thereby the higher the level of the privacy guaranteed.

Infinite Divisibility of Laplace Distribution. In smart grid communications, in order to resist the adversaries against the privacy of participating users, shares of random noise could be generated by each user in the distributed manner. Laplace noise is added probabilistically by each user to achieve (ε, δ)-DP in [13]. Optionally, the infinite divisibility of the Laplace distribution is utilized in [23] to achieve ε-DP. It is proved that in [26] that the Laplace distribution can be assembled from adding i.i.d. gamma distributions. Specifically, suppose $Lap(\lambda)$ denote the random variables having Laplace distribution with $PDFf(x, \lambda) = \frac{1}{2\lambda}e^{-\frac{|x|}{\lambda}}$. Then the infinitely divisible property of the distribution of $Lap(\lambda)$ is obtained. In other words, $Lap(\lambda) = \sum_{i=1}^{n} [G_1(n, \lambda) - G_2(n, \lambda)]$ holds for every integer $n \geq 1$, where $G_1(n, \lambda)$ and $G_2(n, \lambda)$ are i.i.d. random variables having gamma distribution with $PDF\ g(x, n, \lambda) = \frac{1/\lambda^{1/n}}{\Gamma(1/n)}x^{\frac{1}{n}-1}e^{-x/\lambda}$, for $x \geq 0$, and $\Gamma(1/n)$ represents the value of Gamma function at $1/n$.

In our scheme, if the number of smart meters is n, each smart meter, say i, adds $G_1(n, \lambda) - G_2(n, \lambda)$ to its measurement m_i before reporting. Then the sum of the reported data is represented by $\sum_{i=1}^{n} m_i + \sum_{i=1}^{n} [G_1(n, \lambda) - G_2(n, \lambda)] = \sum_{i=1}^{n} m_i + Lap(\lambda)$. Therefore, ε-differential privacy can be provided.

7.4 Proposed Basic DPAFT

In this section, the basic DPAFT for smart grid communications is presented. It mainly includes the following six phases: *system initialization, data aggregation request, data aggregation request relay, user report generation, privacy-preserving report aggregation*, and *secure report reading*. The basic DPAFT mainly focuses on providing fault tolerance for smart metering, which is robust and efficient to handle general failures of measurement report and aggregation.

7.4.1 System Initialization

The single trusted authority (TA) can bootstrap the whole system in the beginning. Concretely, for the system initialization, given the security parameters τ, the *TA* first generate the tuple $(p, q, \mathbb{G})$. Then the TA builds up the BGN PKE, acquires the tuple $(N, \mathbb{G}, g, h)$, where $N = pq$, $g \in \mathbb{G}$ is a random generator of $\mathbb{G}$, and $h = g^{q\beta}$ (for some private β) is a random generator of the subgroup of $\mathbb{G}$ of order p. Note that, this version of BGN PKE is not built upon the bilinear pairing $e : \mathbb{G} \times \mathbb{G} \rightarrow \mathbb{G}_T$. Finally, the *TA* publishes $(N, \mathbb{G}, g, h)$ as the public keys of our system. Although some private keys may also be assigned to users for the sake of authentication, since it is not our main focus, we do not consider it in this work.

In addition, via performing the following steps, the key materials of the residential users $U = \{U_1, U_2, \cdots, U_n\}$ and the CC are assigned by the *TA*.

Step 1: For each user $U_i \in U$, the *TA* first chooses a random number $s_i \in \mathbb{Z}_N$ and assigns s_i as $U_i's$ private key.

Step 2: The TA computes $s_0 \in \mathbb{Z}_N$ such that $s_0 \cdot (s_1 + s_2 + \cdots + s_n) = 1 \ mod \ p$.

Step 3: The TA assigns s_0 as the CC's private key.

Step 4: The TA computes Y_i, such that $Y_i = h^{s_0 s_i}$, for $i = 1, \cdots, n$, and assigns Y_i as the $CC's$ private key as well.

7.4.2 Data Aggregation Request

Assume the reporting time points of our system, *e.g.* every 15 min, are defined as $T = \{t_1, t_2, \cdots, t_{max}\}$ for a sufficient long runtime period. By collecting and analyzing nearly realtime electricity usage data at these time points from all the residential users in the RA, the CC can monitor the health of the whole smart grid system and further provide various high-quality services. Specifically, at time point t_γ, the CC launches a *request* for collecting the usage data in the RA as follows:

Step 1: The *CC* first selects a random $r \in \mathbb{Z}_N$, and computes $A_1 = g^r$ and $A_2 = h^{s_0 r}$.

Step 2: Then, the *CC* sends A_1 and A_2 to the *GW*.

7.4.3 Data Aggregation Request Relay

After receiving A_1 and A_2 from the CC, the *GW* performs the following steps to relay the data aggregation request.

Step 1: The *GW* first selects a random $t \in \mathbb{Z}_N$, and computes $A_3 = A_1^t = g^{rt}$ and $A_4 = A_2^t = h^{s_0 rt}$.

Step 2: Then, the *GW* sends A_3 and A_4 to each $U_i \in U$, respectively.

7.4.4 User Report Generation

Each user $U_i \in U$ collects its usage data $m_i \in \{0, 1, \cdots, W\}$ at time point t_γ, and performs the following steps:

Step 1: U_i first computes the encryption of usage data m_i, using $U_i' s$ private key s_i, as $C_i = A_3^{m_i} A_4^{s_i} = g^{rtm_i} h^{rts_0 s_i}$.

Step 2: Then, U_i reports C_i to the *GW*.

Note that for residential users, the electricity usage within 15 min could not be extremely high, thus choosing an appropriate large W is enough in reality.

7.4.5 Privacy-Preserving Report Aggregation

If all of the n smart meters work correctly, after receiving total n encrypted measurements C_i, for $i = 1, 2, \cdots, n$, at time point t_γ, the GW executes the following procedures for privacy-preserving report aggregation.

Step 1: The *GW* first aggregates the received ciphertexts C_i, for $i = 1, 2, \cdots, n$, as

$$\begin{aligned} C_{\gamma 1} &= (\prod_{i=1}^{n} C_i)^{t^{-1}} \\ &= (g^{r \sum_{i=1}^{n} m_i} h^{rs_0 \sum_{i=1}^{n} s_i})^{tt^{-1}} \\ &= g^{r \sum_{i=1}^{n} m_i} h^r. \end{aligned}$$

Step 2: Then, the *GW* reports the aggregated and encrypted data $C_{\gamma 1}$ at time point t_γ to the *CC*.

Note that, if the smart meters of some users $\hat{U} \subset U$ do not work, *i.e.*, $\hat{U}$ will not report their data at time point t_γ, then the *GW* aggregates the received ciphertexts C_i, for $U_i \in U/\hat{U}$, as

$$\widehat{C_\gamma} = (\prod_{U_i \in U/\hat{U}} C_i)^{t^{-1}}$$
$$= g^{r \sum_{U_i \in U/\hat{U}} m_i} h^{rs_0 \sum_{U_i \in U/\hat{U}} s_i},$$

and sends $\widehat{C_\gamma}$ and $\hat{U}$ to the CC.

7.4.6 Secure Report Reading

If all the n smart meters work correctly, after receiving $C_{\gamma 1}$ from the GW, at time point t_γ, the CC performs the following steps to decrypt the aggregated data.

Step 1: The CC computes $C_{\gamma 1} h^{-r} = g^{r \sum_{i=1}^{n} m_i} = \hat{g}^{\sum_{i=i}^{n} m_i}$, where $\hat{g} = g^r$.

Step 2: Since each $m_i \in \{1, 2, \cdots, W\}$, the sum of all the m_is , *i.e.*, $\sum_{i=1}^{n} m_i \leq nW$, is still not too large. Thus, by computing the discrete log of $\hat{g}^{\sum_{i=1}^{n} m_i}$, the CC can get the sum of users' data $M_{sum} = \sum_{i=i}^{n} m_i$ in expected time $O(\sqrt{n}W)$ using Pollard's lambda method [27].

Note that, if the smart meters of some users $\hat{U} \subset U$ do not work, $\widehat{C_\gamma}$ and $\hat{U}$ will be received at the CC.

Then the CC performs the following procedures to recover the aggregated data $\sum_{U_i \in U/\hat{U}} m_i$,

Step 1: The CC computes

$$\overline{C_\gamma} = \prod_{U_i \in \hat{U}} Y_i$$
$$= h^{s_0 \sum_{U_i \in \hat{U}} s_i}$$

Step 2: The CC computes

$$C_\gamma = \widehat{C_\gamma} (\overline{C_\gamma})^r$$
$$= g^{r \sum_{U_i \in U/\hat{U}} m_i} h^{rs_0 \sum_{U_i \in U/\hat{U}} s_i} h^{rs_0 \sum_{U_i \in \hat{U}} s_i}$$
$$= g^{r \sum_{U_i \in U/\hat{U}} m_i} h^{rs_0 \sum_{i=1}^{n} s_i}$$
$$= g^{r \sum_{U_i \in U/\hat{U}} m_i} h^r.$$

Step 3: Similar as the corresponding procedures when all the smart meters work correctly, the aggregated data $\sum_{U_i \in U/\hat{U}} m_i$ can be recovered successfully.

7.5 The Enhanced DPAFT

In this section we propose the enhanced DPAFT, which provides additional protection of users' privacy against the attack, namely differential attack [20]. In the basic DPAFT, although users' encrypted data are aggregated at the *GW*, so that the individual electricity usage won't be disclosed by the CC or the adversary, through analyzing the aggregated data, their data are still vulnerable to the differential attack which violates users' privacy.

In the enhanced DPAFT, the noise is added to the individual measurement and encrypted by each residential user such that the *GW* can only obtain the noisy encryption. In other words, instead of relying on any centralized party, the noise is generated by the fully distributed means in our proposal. Specifically, inspired by the idea of Infinite Divisibility of Laplace Distribution introduced in Sect. 7.3, each user i, after randomly choosing an appropriate noise, perturbs the smart meter data by simply integrating the encrypted noise firstly. Then, in case that some smart meters do not work correctly, the *GW* complements the noises of the malfunctioning smart meters. In the end, ε-differential privacy is achieved through the collaboration of all the smart meters and the GW in the distributed manner.

The *system initialization* phase, *data aggregation request* phase, and *data aggregation request relay* phase of enhanced DPAFT are exactly the same as those of basic DPAFT, thus we start from describing the *user report generation* phase.

7.5.1 User Report Generation

Each user $U_i \in U$, after collecting its usage data $m_i \in \{0, 1, \cdots, W\}$ at time point t_γ, performs the following steps to report the encrypted and noisy usage data:

Step 1: U_i first calculates value $\widetilde{C_i} = A_3^{m_i+G_1(n,\lambda)-G_2(n,\lambda)} A_4^{s_i} = g^{rt(m_i+G_1(n,\lambda)-G_2(n,\lambda))} h^{rts_0 s_i}$, where $G_1(n, \lambda)$ and $G_2(n, \lambda)$ represent two random values independently sampled from the same gamma distribution, *i.e.*, $G_1(n, \lambda)$ and $G_2(n, \lambda)$ are i.i.d. random variables having gamma distribution with *PDF* $g(x, n, \lambda) = \frac{1/\lambda^{1/n}}{\Gamma(1/n)} x^{\frac{1}{n}-1} e^{-x/\lambda}$, where $x \geq 0$, and n is the number of all the smart meters.

Step 2: U_i then reports $\widetilde{C_i}$ to the *GW*.

7.5.2 Privacy-Preserving Report Aggregation

If all the n smart meters work correctly, after receiving the whole n encrypted and noisy measurements $\widetilde{C_i}$, for $i = 1, 2, \cdots, n$, at time point t_γ, the *GW* executes the following procedures for privacy-preserving report aggregation.

Step 1: The *GW* first aggregates the encrypted and noisy measurements of all the users as

$$\begin{aligned}\widetilde{C_{\gamma 1}} &= (\prod_{i=1}^{n} \tilde{C}_i)^{t^{-1}} \\ &= (g^{r\sum_{i=1}^{n}(m_i+G_1(n,\lambda)-G_2(n,\lambda))} h^{rs_0 \sum_{i=1}^{n} s_i})^{tt^{-1}} \\ &= g^{r(\sum_{i=1}^{n} m_i + Lap(\lambda))} h^{r}.\end{aligned}$$

Step 2: Then, the *GW* reports the aggregated and encrypted noisy data $\widetilde{C_{\gamma 1}}$ at time point t_γ to the *CC*.

Suppose M smart meters $\hat{U} \subset U$ do not work, *i.e.*, $\hat{U}$ will not report their data at time point t_γ, then the *GW* performs the following steps to aggregate the reported data:

Step 1: The *GW* first aggregates the received encrypted and noisy data as

$$\begin{aligned}\widetilde{\widetilde{C_{\gamma 1}}} &= (\prod_{U_i \in U/\hat{U}} \widetilde{C}_i)^{t^{-1}} \\ &= g^{r\sum_{U_i \in U/\hat{U}}(m_i+G_1(N,\lambda)-G_2(N,\lambda))} h^{rs_0 \sum_{U_i \in U/\hat{U}} s_i}\end{aligned}$$

Step 2: The *GW* then complements the M noises of the malfunctioning smarter meters as

$$\begin{aligned}\widetilde{\widetilde{C_\gamma}} &= \widetilde{\widetilde{C_{\gamma 1}}} \prod_{i=1}^{M} A_1^{G_{1i}(n,\lambda)-G_{2i}(n,\lambda)} \\ &= g^{r\sum_{U_i \in U/\hat{U}}(m_i+G_1(N,\lambda)-G_2(N,\lambda))} \\ &\quad \cdot g^{r\sum_{i=1}^{M}(G_{1i}(n,\lambda)-G_{2i}(n,\lambda))} h^{rs_0 \sum_{U_i \in U/\hat{U}} s_i} \\ &= g^{r((\sum_{U_i \in U/\hat{U}} m_i)+Lap(\lambda))} h^{rs_0 \sum_{U_i \in U/U} s_i},\end{aligned}$$

where $G_{1i}(n,\lambda)$ and $G_{2i}(n,\lambda)$, for $i = 1, \cdots, M$, are M pairs of random values independently sampled from the same gamma distribution, *i.e.*, $G_{1i}(n,\lambda)$ and $G_{2i}(n,\lambda)$ are i.i.d. random variables having gamma distribution with *PDF* $g(x,n,\lambda) = \frac{1/\lambda^{1/n}}{\Gamma(1/n)} x^{\frac{1}{n}-1} e^{-x/\lambda}$, where $x \geq 0$.

Step 3: The *GW* finally sends $\widetilde{\widetilde{C_\gamma}}$ and $\hat{U}$ to the *CC*.

7.5.3 Secure Report Reading

If all the n smart meters work correctly, the *secure report reading* phase of enhanced DPAFT is almost the same as that of basic DPAFT. The only difference is that the output the CC gets is no longer the exact one, but the aggregation with Laplace noise, which achieves ε-differential privacy. Specifically, $\sum_{i=1}^{n} m_i + Lap(\lambda)$ is computed from $\widetilde{C_{\gamma 1}} h^{-r} = g^{r(\sum_{i=1}^{n} m_i + Lap(\lambda))}$.

If the smart meters of some users $\hat{U} \subset U$ do not work, $\widetilde{\widetilde{C_\gamma}}$ and $\hat{U}$ will be received by the CC. Then the CC performs the following procedures to recover the aggregation of the noisy measurements of the functioning smart meters,

Step 1: The CC computes

$$\begin{aligned} \overline{C_\gamma} &= \prod_{U_i \in \hat{U}} Y_i \\ &= h^{s_0 \sum_{U_i \in \hat{U}} s_i} \end{aligned}$$

Step 2: The CC computes

$$\begin{aligned} C'_\gamma &= \widetilde{\widetilde{C_\gamma}} (\overline{C_\gamma})^r \\ &= g^{r((\sum_{U_i \in U/\hat{U}} m_i) + Lap(\lambda))} h^{r s_0 \sum_{U_i \in U/\hat{U}} s_i} h^{r s_0 \sum_{U_i \in \hat{U}} s_i} \\ &= g^{r((\sum_{U_i \in U/\hat{U}} m_i) + Lap(\lambda))} h^{r s_0 \sum_{i=1}^{n} s_i} \\ &= g^{r((\sum_{U_i \in U/\hat{U}} m_i) + Lap(\lambda))} h^{r}. \end{aligned}$$

Step 3: Similar as the corresponding procedures when all the smart meters work correctly, the aggregation of the noisy measurements of the functioning smart meters can be recovered as $\sum_{U_i \in U/\hat{U}} m_i + Lap(\lambda)$ successfully.

7.6 Security Analysis

In this section, by security analysis we will show that the proposed DPAFT achieves all the security requirements defined in Sect. 7.2.

7.6.1 Secure Against Eavesdropping Attack

Firstly, an adversary $\mathscr{A}$ may reside in the residential area to eavesdrop the communication flows from the users to the GW. Suppose $\mathscr{A}$ has eavesdropped a ciphertext C_i of user U_i at time point t_γ, *i.e.*, $C_i = A_3^{m_i} A_4^{s_i} = A_3^{m_i} h^{r t s_0 s_i}$. Since the electricity usage m_i within 15 min is usually a small value, the adversary $\mathscr{A}$ may try to launch the brute-force attack by exhaustedly testing each possible value of m_i. However, since the BGN PKE is semantic secure against the chosen ciphertext attack, the adversary $\mathscr{A}$ is not able to recover $U_i's$ usage data m_i without knowing $u_i's$ private key s_i.

Similarly, the aggregated ciphertext, $C_{\gamma 1} = g^{r\sum_{i=1}^{n} m_i} h^r = A_1^{\sum_{i=1}^{n} m_i} h^r$, has the same form as the single user's ciphertext $C_i = A_3^{m_i} h^{rts_0 s_i}$. Since the value r, which is kept secretly by the CC, cannot be obtained by the adversary $\mathscr{A}$, who after eavesdropping the communication flow $C_{\gamma 1}$ from the GW to the CC, cannot obtain the sum of all users' usage data $\sum_{i=1}^{n} m_i$, either, not to mention each user's usage data m_i.

Finally, due to the same reason, when some smart meters, say $\hat{U} \subset U$, are malfunctioning, anyone except the CC cannot recover the sum of functioning users' usage data $\sum_{U_i \in U/\hat{U}} m_i$, let alone each user's private usage data m_i, even if the communication flow of $\widehat{C_\gamma} = g^{r\sum_{U_i \in U/\hat{U}} m_i} h^{rs_0 \sum_{U_i \in U/\hat{U}} s_i}$ and $\hat{U}$ are eavesdropped.

7.6.2 Secure Against Malwares Attack

Even though the adversary $\mathscr{A}$, after deploying some undetectable malwares into the GW and intruding into the database of the GW, has stolen the stored data successfully, he could only get the ciphertexts of all users and the aggregated one. Since the GW does not decrypt any user's measurements, the adversary $\mathscr{A}$ still cannot get any user's private usage data.

In addition, the adversary could also intrude into the database of the CC, but after decryption, the outputs the CC generated are all aggregations of users' measurements, which do not leak individual user's private usage data at all. Therefore, the individual user's report is protected from malwares attack.

7.6.3 Secure in Honest-but-Curious Model

There are totally two possible attack scenes under *honest-but-curious* security model in our scheme. One is the communication flows from the residential users to the GW, which are eavesdropped deliberately and kept improperly by the curious insider participant of the CC or the residential users other than the one reported the usage data. The other is the communication flows from the GW to the CC, which are eavesdropped and kept by the insider participants of residential users improperly. In the former case, because the residential $U_i's$ private key s_i, which cannot be obtained by the CC or other residential user U_j, for $j \neq i$, the CC or the user U_j cannot retrieve the corresponding private usage data m_i from the ciphertext $C_i = A_3^{m_i} A_4^{s_i} = g^{rtm_i} h^{rts_0 s_i}$ reported by the residential user U_i. In the latter case, because the CC's private key r is required to retrieve the aggregated value of $\sum_{i=i}^{n} m_i$, such that $C_{\gamma 1} h^{-r} = g^{r\sum_{i=1}^{n} m_i} = \hat{g}^{\sum_{i=i}^{n} m_i}$, where $\hat{g} = g^r$, any other participant cannot decrypt the sum of users' usage data without processing the secret key r of the CC, not to mention each user's usage data m_i. Due to the same reason, the sum of functioning users' usage data cannot be retrieved by any other participant except for the CC when some smart meters fail to report the measurements, let alone each user's individual usage data.

Therefore, through eavesdropping and analyzing others' communication flows which have nothing to do with oneself, any participant running in the *honest-but-curious* adversary model cannot infer useful knowledge which may help to reveal residential users' privacy.

7.6.4 Secure and Reliable with Fault-Tolerance

We innovate a new approach to realize fault-tolerance of users' malfunction. Even under the circumstances when $\hat{U} \subset U$ do not work, the Y_is, corresponding to the malfunctioning smart meters, which are kept as the secret information by the CC still can be used to recover the aggregated data of $\sum_{U_i \in U/\hat{U}} m_i$. Specifically, the CC, after receiving $\widehat{C_\gamma}$ and $\hat{U}$, uses the private information of Y_is, for $U_i \in \hat{U}$, and r, computes $\overline{C_\gamma} = \prod_{U_i \in \hat{U}} Y_i = h^{s_0 \sum_{U_i \in \hat{U}} s_i}$ and $\widehat{C_\gamma}(\overline{C_\gamma})^r = g^{r \sum_{U_i \in U/\hat{U}} m_i} h^r$ to recover the sum usage of functioning smart meters, *i.e.*, $\sum_{U_i \in U/\hat{U}} m_i$. Because Y_is and r are kept secret by the CC, without them, anyone else cannot recover the sum of functioning smart meters' usage data, not to mention each user's private usage data m_i.

7.6.5 Secure Against Differential Attack

Although, as abovementioned, the user's electricity usage privacy is protected from eavesdropping, malwares attack, *etc.*, the adversary $\mathscr{A}$ still can launch some differential attacks to threaten user's privacy if he gets aggregations of two adjacent data sets. In our enhanced scheme, all the functioning smart meters and the *GW* collaboratively add appropriate Laplace noise in the form of ciphertext, which achieves ε-differential privacy. Specifically, $\sum_{i=1}^{n} m_i + Lap(\lambda)$ and $\sum_{U_i \in U/\hat{U}} m_i + Lap(\lambda)$ are computed by the CC when all the n smart meters work correctly and M (the size of $\hat{U}$) malfunctioning smart meters occur, respectively. Therefore, even though the adversary $\mathscr{A}$ obtains aggregations of two adjacent data sets, the difference of the two aggregation results still does not leak the individual user's private data to $\mathscr{A}$ at all.

7.7 Performance Evaluation

In this section, we evaluate the performance of the proposed DPAFT in terms of *storage cost*, *computation complexity*, *utility of differential privacy*, *robustness of fault tolerance*, and *efficiency of user addition and removal*. Because few of the existing schemes supports fault tolerance and differential privacy simultaneously, in this section, we compare our proposed scheme with the state of the art schemes [13, 23, 24, 28], which support privacy-preserving aggregation with fault tolerance or differential privacy.

7.7.1 Storage Cost

In the scheme of [24], it is necessary for the *GW* to configure the huge amount of memory buffers to store the *future ciphertexts* for all the residential users. Even though such overhead is acceptable when the scale of the scheme is not large as pointed out by the authors in [24], when the scale of the scheme turns large, *e.g.*, with millions of smart meters, it brings considerable storage overhead to the *GW*. By contrast, in our scheme, the *GW* is just responsible for data aggregation and packages relay, thus there is no special storage requirements, which makes it more rational and practical.

7.7.2 Computation Complexity

In the scheme of [24], each user should select k other users as *partners* to encrypt the measurements. Specifically, in the *initial setup* phase of their scheme, the shared secret keys between every two users of the *partner pairs* should be generated and assigned secretly. Then, in *data report phase*, the two parts of ciphertexts, *i.e.*, the *current ciphertext* and the *future ciphertext*, should be calculated and reported. Each user generates the *current ciphertext* by adding the random number and the noise information to the real measurement. The random number is computed by using the shared secret key assigned in *initial setup* phase and the information of reporting time point. Then, the *future ciphertext* should also be generated and reported simultaneously for achieving fault tolerance. In our proposed scheme, each user independently report the measurement, thus there is no need to compute and assign the shared secret keys among the users. And the additional computation of *future ciphertext* is not necessary either. Our DPAFT only needs two exponentiation operations and one multiplication operation totally for reporting the measurement of each user. The computation complexity is less than or at least not heavier that the scheme of [24]. Because our scheme is secure in *honest-but-curious* adversary model as illustrated in aforementioned Sect. 7.6, while the scheme of [24] is not constructed in this security model, other comparison of computation complexity for the *GW* and the *CC* cannot be carried out. Nevertheless, all the computation cost of the *GW* and the *CC* include only a handful of exponentiation, add, and multiplication operations, which leads to the low computation complexity of our scheme.

Similar as [13], since the techniques to support data aggregation of our scheme are also mainly inspired by the BGN PKE [25], and the scheme of [13] is the most efficient one based on BGN PKE, we also compare our scheme with [13] in terms of computation complexity. In the most time-consuming phase of *secure report reading*, both of our scheme and [13] take two exponentiation operations and one multiplication operation. Additionally, in order to support fault tolerance, for each round of data aggregation, our scheme costs two exponentiation operations and one multiplication operation in *data aggregation request* phase, and two exponentiation operations in *data aggregation request relay* phase, respectively. Since all these operations are performed by the entities of the CC or the GW, and both of which are with powerful computational capability, thus, this increases negligible computation overhead. Therefore, with additional negligible computational cost, our scheme achieves fault tolerance while the scheme of [13] cannot support.

7.7.3 Utility of Differential Privacy

Similar to the scheme of [24], we implement an electricity consumption simulator having the ability of generating realistic 1-min consumption traces synthetically. It is extended from the basic simulator presented in [29]. We produce traces for 2000 households based on this simulator. Specifically, the distribution of the residents of each household consults the UK statistics on household sizes in 2011 [30]. We select the day to be a weekday in February. For the appliances in a household, we choose them randomly among 33 available ones. For differential privacy, ε is set to 1, and the global sensitivity is set to 33 kW, which is the sum of power demands of all the appliances and lights.

In the scheme of [24], in order to resist the attack from differentiating the *current ciphertext* and the *future ciphertext*, an additional Laplace noise $Lap(\lambda)$ is added to each smart meter's *future ciphertext*. However, as claimed by the authors in [24], this incurs large errors which increases greatly with the increasing number of malfunctioning smart meters. Specifically, if w smart meters failed, $w+1$ values of Laplace noise will be added additionally to the *future ciphertexts*, which generates $O(\sqrt{w+1})$ error as explained in [24]. Our scheme overcomes this drawback, thus, it is of better utility than the scheme of [24] as compared in Fig. 7.3. The figures illustrate the traces of the actual total measurements, the noisy total consumptions of the scheme [24] and ours, respectively, for the different parameters, where in each of the figure, n and p denote the total number of the household, and the different ratio of malfunctioning smart meters, respectively. As it can be seen from the figures, the larger the number of p, the more accurate of our scheme comparing with the scheme of [24] in terms of the utility of differential privacy. Let the 1-h RMSE (root mean square error) of Jongho et al.'s protocol [24] and DPAFT be γ_1 and γ_2, respectively. The ratio of $\gamma = \frac{\gamma_1}{\gamma_2}$ with p is depicted in Fig. 7.4a.

More precisely, we measure the closeness between the sequences of actual and noisy sums by the 1-day *RMSE*, *i.e.*, $RMSE = \sqrt{(\frac{1}{T} \cdot \sum_{t=1}^{T} (\widehat{m}_t - m_t)^2)}$, where $T = 1440$ is the number of time points of 1 day, $\widehat{m}$ is the noisy sum, and m is the real sum. Figure 7.4 compares the 1-day *RMSE* of our scheme and that of the scheme of [24].

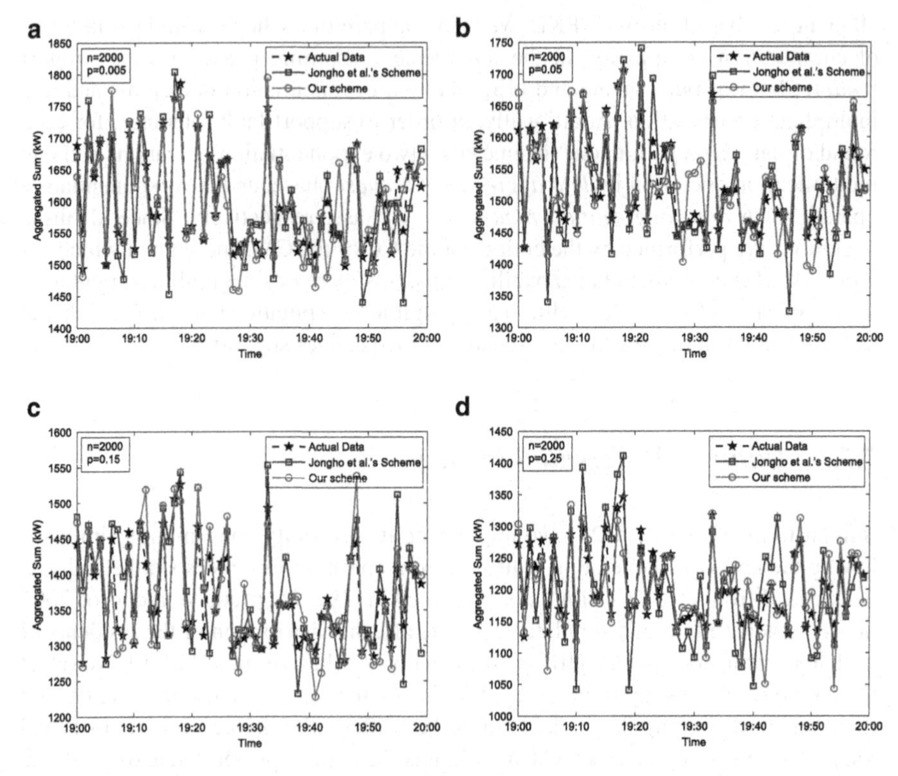

Fig. 7.3 Comparison of noisy total consumption between DPAFT and Jongho et al.'s protocol [24]. (**a**) $n = 2000, p = 0.005$; (**b**) $n = 2000, p = 0.05$; (**c**) $n = 2000, p = 0.15$; (**d**) $n = 2000, p = 0.25$

Although the authors of [24] claimed that their scheme outperforms the existing scheme [28], the RMSE of [24] increases inconspicuously only in the scenarios where the failure ratio of p is very small, *i.e.*, with the maximal magnitude of about 0.001 as illustrated in their scheme [24]. We take into account the more reasonable scenarios as most of the literatures assumed [13, 23], where the magnitude of p can be much larger than that of the scheme [24] considered. In such scenarios, the effects of the additional Laplace noises $Lap(\lambda)$ added to the *future ciphertexts* of [24] cannot be neglected. Our scheme improves accuracy for large scale of report failures in smart grid communications mainly because of the elimination of these noises.

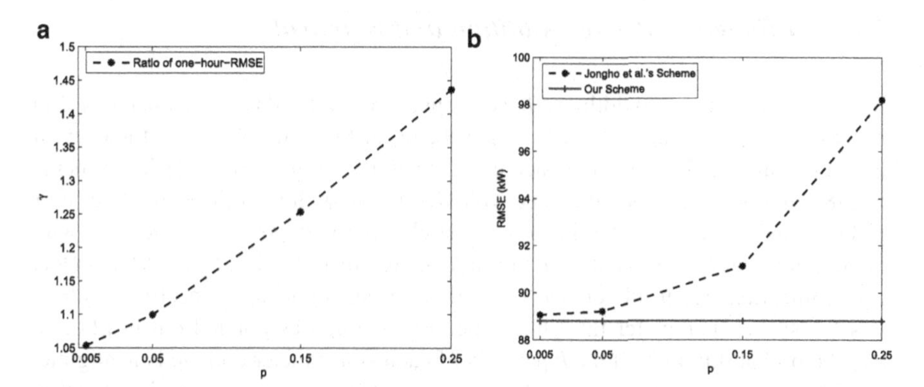

Fig. 7.4 Comparison of (**a**) 1-h RMSE and (**b**) 1-day RMSE between DPAFT and Jongho et al.'s protocol [24]

7.7.4 Robustness of Fault Tolerance

When some meters fail to report, the smart meter system still needs to be able to aggregate the measurements of the remaining functioning ones to perform real-time data monitoring and analyzing successfully. In the scheme of [24], the GW stores B pieces of *future ciphertexts* for each smart meter to support fault tolerance. Without loss of generality, assume the data report interval of the smart meter system of [24] is T. And suppose at time point T_a, due to some fault, certain smart meter U_i cannot report the measurement successfully to the GW, and the fault recovery time point is T_b. Thus, the fault duration period T_{per} is $T_b - T_a$. If $T_{per} > B \cdot T$, the system of [24] cannot tolerate the fault any longer after the time point of $T_a + B \cdot T$, because the pre-stored *future ciphertexts* are used up, until the fault smart meter U_i is to be recovered again at T_b. The robustness of fault tolerance turns to be much worse when the number of the malfunctioning smart meters increases. In order to support more robust fault tolerance, the system parameter of the buffer size B of [24] should be increased further. However, this causes heavy storage cost, computation complexity and communication overhead as illustrated in the aforementioned performance analysis. By contrast, our scheme is more robust of fault tolerance and can support data aggregation with any rational number of malfunctioning smart meters with arbitrary long fault period, because the mechanism of our fault tolerance is not related to the malfunctioning smart meters directly and is independent of any external factors, *e.g., future ciphertexts*.

7.7.5 *Efficiency of User Addition and Removal*

In [13], both of the bandwidth overhead for user addition and removal are $O(N)$, and it does not support fault tolerance. The scheme of [23] improves the efficiency of user removal to $O(1)$, while only achieving *partial fault tolerance* [24]. The scheme of [28] supports fault tolerance but with $O(N)$ bandwidth overhead for both user addition and removal. The scheme of [24] claims to support fault tolerance with highest efficiency in terms of user addition and removal. It takes $O(k \times B)$ and $O(k)$ communication overhead for one user addition and removal, respectively, where k is the system parameter that should be large enough as pointed out in [24], so that the probability that all the k partners of each smart meter collude is negligible. In addition, when some user, say U_i, decides to leaves the system taking effect at time point t, U_i will have to inform the aggregator of its leave decision in advance at time point $t - B$, and it is also necessary for the aggregator to broadcast the leave announcement message containing the *id* of U_i and t, denoted by $leave(i, t)$. This incurs additional communication overhead and causes delays for the effect of user addition or removal. The assumption that such delays is acceptable is generally unreasonable, because in most circumstances, it is unpredictable of the dynamics of users.

Our scheme is more efficient than the scheme of [24] in terms of user addition and removal. In our scheme, to support user addition, the *TA* just needs to reassign the key materials for the newly added users and the *CC*, and to support user removal, the *TA* just needs to update the key materials for the *CC*. There is no need of any other operation for the other participants in the system. Specifically, the *TA* just needs to recompute s_0' and s_i' corresponding to the changed users, such that $s_0' \cdot (\sum_{U_i \in U_{unchanged}} s_i + \sum_{U_i' \in U_{changed}} s_i') = 1 \ mod \ p$, reassign the corresponding key material s_i' to each of the changed user, and reassign Y_i' related to each of the changed user and s_0' to the *CC* only. Moreover, in our proposed scheme, it takes nearly no time for the changed users to take effect. Therefore, our scheme is most efficient and supports the real meaning of *full-dynamic* of user addition and removal.

7.8 Related Work

In this section, we briefly discuss some other research works [2, 12–14, 22–24], which are closely related to our scheme. In [13], Shi et al. propose an aggregation scheme that can protect individual user's privacy and is secure against differential attack. However, it is unclear how to extend their scheme to be fault tolerant. Furthermore, only the relaxed (ε, δ)-differential privacy definition is obtained, which diminishes the utility [24]. In [2], Lu et al. propose a protocol that supports multidimensional data aggregation (i.e., PPMDA in Chap. 3). The protocol significantly reduces the computation and communication overhead by encrypting multi-dimensional data into one single ciphertext, which can also be applied in our proposed scheme. But their scheme is not secure under a more rational security model of *honest-but-curious* considered in this study. In [14], Jia et al. propose a privacy-preserving data aggregation scheme, in which coefficients of the polynomial are used to hide users' individual measurements. In [22], Chen et al. propose one privacy-preserving multi-functional data aggregation scheme for smart grid communications achieving privacy-preserving aggregation of multiple functions such as average, variance, one-way ANOVA, *etc*. They also extend their scheme to be secure against differential attacks upon multi-functional aggregations. However, it is unclear how to extend the aforementioned two schemes [14, 22] to support fault tolerance. In [12], Chen et al. propose a privacy-preserving data aggregation scheme with fault tolerance. Their scheme provides privacy-preserving aggregation against a strong adversary which may compromise a few servers at the control center, and supports fault tolerance of residential users and servers. However, although users' encrypted data are aggregated so that the individual electricity usage won't be disclosed, their data are still vulnerable to the differential attack which violates user's privacy.

In [23], Acs and Castelluccia proposed a privacy-preserving smart metering system which is secure against differential attack by adding Laplace noise in the distributed manner. However, the scheme requires the establishment of pairwise keys between each pair of smart meters, which increases the required network bandwidth and delay. They also extend their basic scheme intended to be fault tolerant. However, the extended scheme only support *partial fault tolerance* [24]. Furthermore, because the extended scheme assumes the fixed maximum number (*i.e.* M) of untrustful smart meters, and adds extra noise needed to ensure differential privacy, the utility is decreased if more than $N - M$ smart meters add the noises faithfully. In [24], Jongho Won et al. propose a fault-tolerant aggregation protocol for privacy-assured smart metering, their protocol introduces *future ciphertexts* to support fault tolerant of possible malfunctioning smart meters, which leads to the heavy round-based communication, computation and storage overhead. Furthermore, the more malfunctioning smart meters are, the larger error the protocol is. Specifically, if w smart meters failed, $w + 1$ values of additional Laplace noise should be added. Therefore, the proposed scheme generates $O(\sqrt{w+1})$ error when w smart meters fail to report.

Although our proposed DPAFT scheme addresses the similar issues, *i.e.* providing efficient, privacy-preserving, and differentially private aggregation in smart grid communications, and supporting fault tolerance, in contrast to the above works, our research emphases still have some differences: (a) we propose our aggregation protocol in a more challenging threaten model which covers external attack, internal attack under *honest-but-curious model*, differential attack, and malware attack; and (b) we take differential privacy and fault tolerance into consideration at the same time, thus it additionally improves the reliability and practicability.

7.9 Summary

In this chapter, a new secure data aggregation scheme, named DPAFT, for smart grid system has been proposed. DPAFT is secure under the more challenging threaten model which covers external attack, internal attack under *honest-but-curious* model, differential attack, and malware attack. Furthermore, differential privacy and fault tolerance are taken into consideration at the same time, thus further improves the reliability and practicability. Unlike those existing similar works, the fault-tolerance approach put forward in this work is based on a novel artful constraint relation $s_0 \cdot \sum_{i=1}^{n} s_i = 1$. With such novel constraint, the fault tolerance is robust, efficient and flexible, and the real meaning of *full-dynamic* of user updates is realized efficiently. Through extensive performance evaluation, we have also demonstrated that DPAFT outperforms the state of the art schemes in terms of storage cost, computation complexity, utility of differential privacy, robustness of fault tolerance, and the efficiency of user addition and removal.

References

1. H. Bao and R. Lu, "A new differentially private data aggregation with fault tolerance for smart grid communications," *IEEE Internet of Things Journal*, vol. 2, no. 3, pp. 248–258, 2015. [Online]. Available: http://dx.doi.org/10.1109/JIOT.2015.2412552
2. R. Lu, X. Liang, X. Li, X. Lin, and X. Shen, "Eppa: An efficient and privacy-preserving aggregation scheme for secure smart grid communications," *IEEE Transactions on Parallel and Distributed Systems*, vol. 23, no. 9, pp. 1621–1631, 2012.
3. X. S. Shen, "Empowering the smart grid with wireless technologies [editor's note]," *IEEE Network*, vol. 26, no. 3, pp. 2–3, 2012.
4. D. Li, Z. Aung, J. Williams, and A. Sanchez, "P3: Privacy preservation protocol for automatic appliance control application in smart grid," *IEEE Internet of Things Journal*, vol. 1, no. 5, pp. 414–429, 2014.
5. D. Banerjee, B. Dong, M. Taghizadeh, and S. Biswas, "Privacy-preserving channel access for internet of things," *IEEE Internet of Things Journal*, vol. 1, no. 5, pp. 430–445, 2014.
6. Y. Wang, S. Mao, and R. Nelms, "Distributed online algorithm for optimal real-time energy distribution in the smart grid," *IEEE Internet of Things Journal*, vol. 1, no. 1, pp. 70–80, 2014.
7. X. Li, X. Liang, R. Lu, X. Shen, X. Lin, and H. Zhu, "Securing smart grid: cyber attacks, countermeasures, and challenges," *IEEE Communications Magazine*, vol. 50, no. 8, pp. 38–45, 2012.
8. Y. Wang, S. Mao, and N. R.M., "Distributed online algorithm for optimal real-time energy distribution in the smart grid," *IEEE Internet of Things Journal*, vol. 1, no. 1, pp. 70–80, 2014.
9. J. Lin, K. Leung, and V. Li, "Optimal scheduling with vehicle-to-grid regulation service," *IEEE Internet of Things Journal*, vol. 1, no. 6, pp. 556–569, 2014.
10. H. Liang, B. J. Choi, A. Abdrabou, W. Zhuang, and X. Shen, "Decentralized economic dispatch in microgrids via heterogeneous wireless networks," *IEEE Journal on Selected Areas in Communications*, vol. 30, no. 6, pp. 1061–1074, 2012.
11. K. Kursawe, G. Danezis, and M. Kohlweiss, "Privacy-friendly aggregation for the smart-grid," in *Privacy Enhancing Technologies*. Springer, 2011, pp. 175–191.
12. L. Chen, R. Lu, and Z. Cao, "Pdaft: A privacy-preserving data aggregation scheme with fault tolerance for smart grid communications," *Peer-to-Peer Networking and Applications*, pp. 1–11, 2014.

13. E. Shi, T.-H. H. Chan, E. G. Rieffel, R. Chow, and D. Song, "Privacy-preserving aggregation of time-series data." in *NDSS*, vol. 2, no. 3, 2011, p. 4.
14. W. Jia, H. Zhu, Z. Cao, X. Dong, and C. Xiao, "Human-factor-aware privacy-preserving aggregation in smart grid," *Systems Journal, IEEE*, vol. 8, no. 2, pp. 598–607, 2014.
15. F. Li, B. Luo, and P. Liu, "Secure information aggregation for smart grids using homomorphic encryption," in *2010 First IEEE International Conference on Smart Grid Communications (SmartGridComm)*. IEEE, 2010, pp. 327–332.
16. I. Stojmenovic, "Machine-to-machine communications with in-network data aggregation, processing, and actuation for large-scale cyber-physical systems," *IEEE Internet of Things Journal*, vol. 1, no. 2, pp. 122–128, 2014.
17. P. Paillier, "Public-key cryptosystems based on composite degree residuosity classes," in *Advances in cryptology—EUROCRYPT'99*. Springer, 1999, pp. 223–238.
18. F. D. Garcia and B. Jacobs, "Privacy-friendly energy-metering via homomorphic encryption," in *Security and Trust Management*. Springer, 2011, pp. 226–238.
19. V. Rastogi and S. Nath, "Differentially private aggregation of distributed time-series with transformation and encryption," in *Proceedings of the 2010 ACM SIGMOD International Conference on Management of data*. ACM, 2010, pp. 735–746.
20. C. Dwork, "Differential privacy," in *Automata, languages and programming*. Springer, 2006, pp. 1–12.
21. ——, "Differential privacy: A survey of results," in *Theory and Applications of Models of Computation*. Springer, 2008, pp. 1–19.
22. L. Chen, R. Lu, Z. Cao, K. AlHarbi, and X. Lin, "Muda: Multifunctional data aggregation in privacy-preserving smart grid communications," *Peer-to-Peer Networking and Applications*, pp. 1–16, 2014.
23. G. Acs and C. Castelluccia, "I have a dream!(differentially private smart metering)," in *Information Hiding*. Springer, 2011, pp. 118–132.
24. J. Won, C. Y. Ma, D. K. Yau, and N. S. Rao, "Proactive fault-tolerant aggregation protocol for privacy-assured smart metering," in *INFOCOM 2014*. IEEE, 2014, pp. 2804–2812.
25. D. Boneh, E.-J. Goh, and K. Nissim, "Evaluating 2-dnf formulas on ciphertexts," in *Theory of cryptography*. Springer, 2005, pp. 325–341.
26. S. Kotz, T. Kozubowski, and K. Podgorski, *The Laplace Distribution and Generalizations: A Revisit With Applications to Communications, Exonomics, Engineering, and Finance*. Springer, 2001, no. 183.
27. A. J. Menezes, P. C. Van Oorschot, and S. A. Vanstone, *Handbook of applied cryptography*. CRC press, 2010.
28. T.-H. H. Chan, E. Shi, and D. Song, "Privacy-preserving stream aggregation with fault tolerance," in *Financial Cryptography and Data Security*. Springer, 2012, pp. 200–214.
29. I. Richardson, M. Thomson, D. Infield, and C. Clifford, "Domestic electricity use: A high-resolution energy demand model," *Energy and Buildings*, vol. 42, no. 10, pp. 1878–1887, 2010.
30. O. for National Statistics, "Families and households, 2001 to 2011," Jan. 2012, http://www.ons.gov.uk/ons/rel/family-demography/families-and-households/2011/rft-tables-1-to-8.xls.

[illegible] R. Lu, [illegible] H. [illegible] Choo, and D. Song, "Privacy-preserving aggregation of [illegible]," in NDSS, vol. 2, no. 3, 2011, p. 4.
[illegible] W. Jia, H. Zhu, Z. Cao, X. Dong, and C. Xiao, "Human-factor-aware privacy-preserving aggregation in smart grid," IEEE Systems Journal, vol. 8, no. 2, pp. 598–607, 2014.
[illegible] F. Li, B. Luo, and P. Liu, "Secure information aggregation for smart grids using homomorphic encryption," in 2010 First IEEE International Conference on Smart Grid Communications. IEEE, 2010, pp. 327–332.
[illegible]
[illegible] Springer, 1998, pp. [illegible]
[illegible] energy [illegible] 2010, pp. 226–[illegible]
[illegible] 2011, pp. 1–[illegible]
[illegible]
[illegible]
[illegible]
[illegible]
[illegible] 2008, pp. 1–[illegible]
[illegible]
[illegible] S. Vaudenay [illegible]
[illegible] Springer, 2011, pp. 140–[illegible]
[illegible]
[illegible] 2017. [illegible]

Chapter 8
Privacy-Preserving Data Aggregation with Data Integrity and Fault Tolerance

In the last two chapters, we have discussed two privacy-preserving data aggregation schemes with fault tolerance for secure smart grid communications. In this chapter, we will introduce a novel secure data aggregation scheme to achieve privacy preservation and data integrity with differential privacy and fault tolerance [1], obtaining a good tradeoff of accuracy and security of differential privacy for arbitrary number of malfunctioning smart meters.

8.1 Introduction

Comparing with the traditional power grid, smart grid has assimilated various technologies, e.g., data communication and analyzing, sophisticated control and sensing technologies, into the traditional power grid, enables the power distribution to be more efficient and reliable from power generation, transmission, and distribution to end user consumption, and supports the renewable energy [2]. Specifically, as illustrated in Fig. 8.1, by deploying and configuring various sensors following the two-way flows of electricity and communication in smart grid, enormous amounts of real-time information is collected and delivered to the control center (CC) for real-time monitoring and analyzing the health of power grid.

However, the real-time user data, e.g., collected every 15 min, contain detailed power usage habits which highly correlates to user's privacy, thus they must be protected from unauthorized accesses. In addition to privacy preservation, data integrity is also critical in smart grid communication, otherwise, an attacker could steal or pollute energy usage and consumption information to diminish the availability of smart grid. Therefore, it is of great significance to simultaneously preserve user privacy and assure data integrity in smart grid communications.

In order to address the privacy issues, several privacy-preserving data aggregation schemes for smart grid communications have been proposed [2–11]. Most of them

R. Lu, *Privacy-Enhancing Aggregation Techniques for Smart Grid Communications*, Wireless Networks, DOI 10.1007/978-3-319-32899-7_8

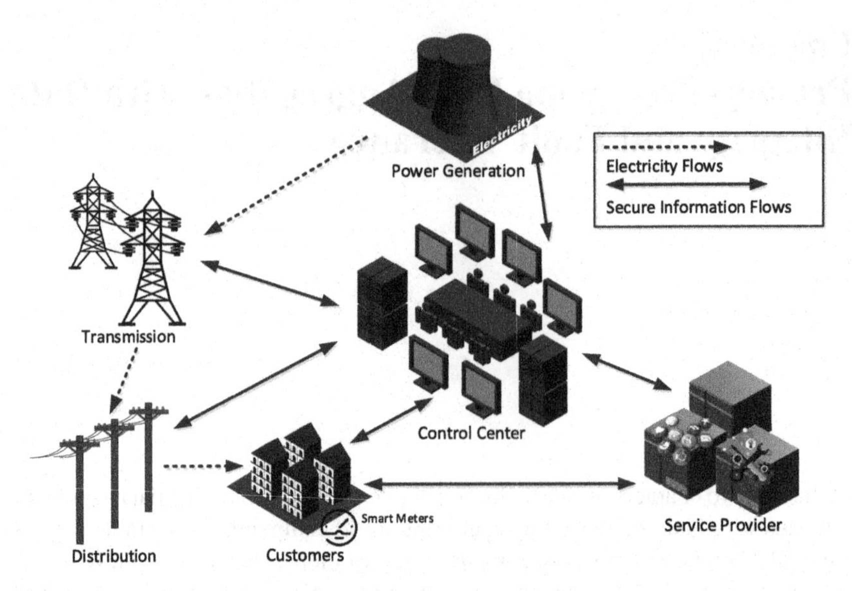

Fig. 8.1 The conceptual smart grid system architecture

[2, 3, 6, 7, 10] utilize the homomorphic encryption techniques [12] to encrypt and aggregate users' data in the local area gateway (GW) and forward the data to CC without decryption. However, they only consider protection user's privacy against GW, while CC is still easy to obtain individual user's data. This is because the private key CC holds may not only be used to decrypt the aggregated data, but also be abused to reveal any single user's electricity usage. More seriously, some strong adversaries may deploy undetectable malwares to GW or CC for privacy disclosure of users [3, 10]. This may also conflict users' privacy concerns. When strong adversaries are considered, who aim to snoop user's privacy, these privacy-preserving data aggregation schemes are not robust enough to keep user's privacy unexposed. Other aggregation schemes like [4, 5, 8, 9, 11] take advantage of key management techniques, i.e., the sum of all participants' (including all users and CC) random numbers equals to 0, to diminish CC's authority. One major drawback of such mechanisms is that they are not able to tolerate the report failures [4, 8, 11]. Even though single user fails to report data at some time point, CC would not be able to get anything because the sum of the random numbers in the ultimate encrypted aggregation is no longer 0. This can be a big problem because smart meters, as low-cost devices running in the unprotected environments, are prone to failures. Another challenging problem that secure data aggregation schemes could face comes from differential attack [13]. Specifically, even though one aggregation scheme is secure, once CC obtains the summation of n users and the counterparts of $n-1$ users, the privacy of the differential one can be inferred to impair user's privacy. This problem has been studied in several works, such as [4, 13–15]. However, most of them do not support fault tolerance, i.e., it is infeasible to be extended to the scenarios where malfunctioning smart meters or communication failures occur. A handful of them [5, 9] support fault tolerance either with unsatisfying utility or with low efficiency, which are not practical, especially in circumstances when the precise number of malfunctioning smart meters are unpredictable and sometimes may be large to some extent.

Meanwhile, in order to prevent malicious adversaries from impair (e.g., modify, forge, injection, reply and/or delay, etc.) user's usage data report, several message authentication schemes [16–21] have also been presented to ensure data integrity in smart grid communications. Generally, the existing techniques for authenticating communications in smart grid mainly include Hash to Obtain Random Subsets Extension (HORSE) [17], Bins Balls (BiBa) [16], Digital Signature Algorithm (DSA) [18] and MAC/HMAC (hash-based message authentication code) [19]. Comparing with HORSE and BiBa, DSA achieves higher security, however, the computational complexity, especially at the user side is still very heavy [22]. Performance evaluations show that the MAC/HMAC based authentication technique is more efficient than DSA [19]. However, public key based session key agreement protocol is needed in each round to ensure data integrity, which still poses heavy computational cost and communication overhead for practical applications.

In addition, in bandwidth-intensive and delay-sensitive smart grid communications, especially in user side, efficient mechanism with low computational cost and communication overhead for smart grid communications should be designed to speed up the real applications.

Thus, how to realize an efficient, secure (with privacy preservation and data integrity simultaneously) data aggregation having enhanced and robust properties (fault tolerant of failures and secure against differentially attack) for smart grid communications still deserves further investigations. In this chapter, we propose a novel secure and lightweight data aggregation scheme achieving privacy preservation and data integrity with differential privacy and fault tolerance for smart grid communications. Specifically, the main contributions of this chapter are four-fold.

- Firstly, by introducing auxiliary ciphertext subtly, we put forward a novel distributed solution for fault tolerant data aggregation. Unlike most of the existing similar works, which depend on the central trust authority to trace and separate the malfunctioning smart meters from the functioning ones to be able to aggregate the smart meter measurements in case of report failures, our proposed scheme supports fault tolerance of malfunctioning smart meters without the participation and restriction of any external factors. Specifically, utilizing the auxiliary ciphertexts, CC can obtain the aggregation of the functioning smart meters flexibly and efficiently for any rational number of malfunctioning smart meters with arbitrary long failure period.
- Secondly, observing the fact that user's private data may often suffer from differential attacks, our proposed scheme provides differential privacy by adding appropriate noises chosen from Symmetric Geometric distribution to the aggregation data by GW. To the best of our knowledge, most of the existing similar works cannot support differential privacy and fault tolerance at the same time.

A handful of literatures trying to address this problem only consider the scenarios that there is small amount (or fixed maximum number) of malfunctioning smart meters to be able to add appropriate noises to support differential privacy. Our scheme supports differential privacy and fault tolerance simultaneously, and achieves a good tradeoff of accuracy and security of differential privacy for arbitrary number of malfunctioning smart meters.

- Thirdly, by integrating a pair of identities and private/public keys of two communication parties, and current time slot for data report, a novel efficient authentication technique is proposed to flexibly generate and share session keys in noninteractive way. The shared session key is leveraged for AES encryption to achieve source authentication and data integrity of transmitted data. The security analysis and performance evaluation indicate that the proposed mechanism can efficiently and effectively prevent the malicious adversary from impairing and polluting (e.g., modify, forge, injection, reply and/or delay, etc.) the transmitted data.
- Finally, through decentralizing the computational overhead and the power of the hub-like entity GW, which is usually with limited computation resources and is semi-trust, the security of our proposed scheme is enhanced and the efficiency is improved significantly. Specifically, only the encryption of the usage data C_i and the auxiliary ciphertext δ_i are aggregated and processed beforehand by at least two users, respectively, can they be reported to GW. In addition, through comparative performance analysis, we demonstrate that our proposed data aggregation scheme outperforms the state-of-the-art similar schemes [4–6] in terms of computation complexity, communication cost, robustness of fault tolerance, and utility of differential privacy.

The remainder of this chapter is organized as follows. We first identify the problem formalization which includes system model, attack model and design goal in Sect. 8.2, and briefly recall some preliminaries in Sect. 8.3. Then, we present our proposed data aggregation scheme in Sect. 8.4. Subsequently, the security analysis and performance evaluation are presented in Sects. 8.5 and 8.6, respectively. We also discuss some related work in Sect. 8.7. Finally, we draw our conclusions in Sect. 8.8.

8.2 Problem Formalization

In this section, we formalize system model, attack model, and identify our design goal.

8.2.1 System Model

Considering real application requirements of smart grid communications, residential users pay great attention to their privacy when reporting measurements to the control center (CC), smart meters as inexpensive home devices, which are often deployed in unprotected environments, may fail to report the usage data, and users' reported data may be tampered with by the malicious adversary due to the unreliable network channel. In this work, we mainly put our emphasis on how to report users' measurements to CC in a secure (with privacy preservation and data integrity simultaneously) and reliable (with fault tolerance) way. Specifically, in our system model, a typical smart grid communication architecture is considered, as shown in Fig. 8.2, which includes a trusted authority (TA), a control center (CC), a residential gateway (GW), and a huge amount of residential users $U = \{U_1, U_2, \ldots, U_n\}$ in a residential area (RA).

Trusted Authority (TA): The TA is a trustable entity who has powerful processing capacity and is in charge of management of the whole system.

Control Center (CC): CC is a highly-trusted entity, whose responsibility is to collect, process and analyze the nearly real-time data to be able to provide reliable services for smart grid.

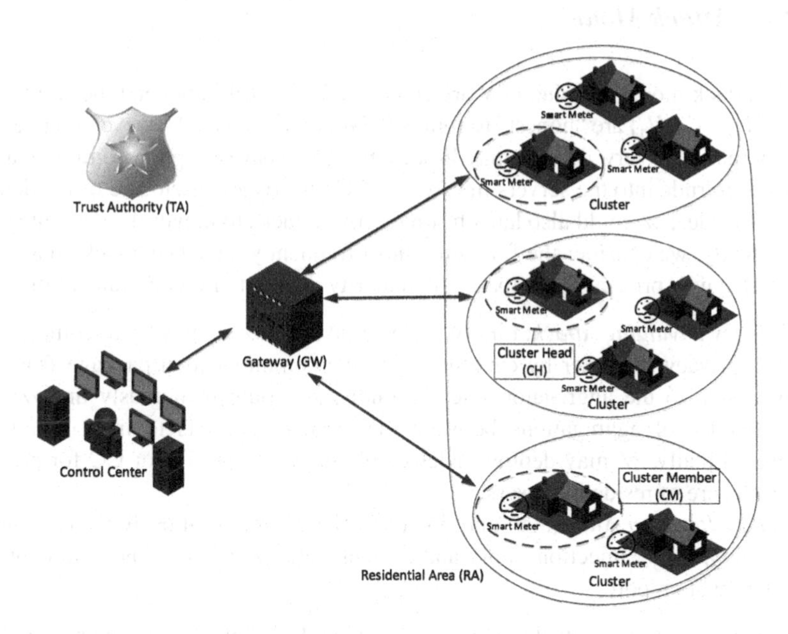

Fig. 8.2 System model under consideration

Gateway (GW): GW mainly performs two functions, i.e., aggregation and relaying. The responsibility of aggregation is to aggregate residential users' measurements into a integrated one, and the responsibility of relaying is to help forward the communication flows between CC and residential users in a secure way.

Residential Users $U = \{U_1, U_2, \ldots, U_n\}$: Each residential user $U_i \in U$ is equipped with various smart appliances and a smart meter to collect the real-time electricity usage data.

Note that, for the sake of high efficiency and security consideration, our protocol stipulates that any residential user U_i cannot report its measurement individually and directly to GW. Specifically, based on the geographical adjacency, e.g., neighboring districts, U is divided into w clusters $\{\mathrm{CL}_1, \cdots, \mathrm{CL}_w\}$, and the number of users in cluster CL_i is n_i. Here, w is a system parameter, which is dependent on the whole topology and the scale of n. In each cluster $\mathrm{CL}_i \subseteq U$, a cluster head (CH), called U_{h_i}, is appointed. All the other users are cluster members (CMs). Actually, U_{h_i} itself is also a CM of CL_i. There is no particular requirement for appointing the unique CH in each cluster, who can be fixed beforehand or randomly selected temporarily. Then, each CH U_{h_i}, for $i = 1, \cdots, w$, is responsible for pre-aggregating and forwarding all the data reported in CL_i to GW in a certain period, e.g., every 15 min.

8.2.2 Attack Model

In our attack model, CC and GW are considered to be trustable, and the residential users $\{U_1, \cdots, U_n\}$ are "honest" to follow the protocol as well. However, there exists an external adversary $\mathscr{A}$ who may lurk in the RA to eavesdrop the communication flows or intrude into the servers in GW and CC for privacy disclosure of residential users. Besides, $\mathscr{A}$ could also launch some active attacks to impair the data integrity. Specifically, we consider the following most frequently launched attacks intending to divulge user privacy and impair data integrity in smart grid communications:

1. *Privacy Divulging Attack*: Firstly, an external attacker $\mathscr{A}$ may try to compromise data privacy of a user by eavesdropping the communication package from the user side to the aggregator side. Secondly, $\mathscr{A}$ could maliciously analyze the difference of aggregations between the similar data sets to infer individual ones. Finally, $\mathscr{A}$ may deploy undetectable malwares to GW or CC for privacy disclosure of residential users.
2. *Data Alteration Attack*: $\mathscr{A}$ may intercept the communication links and impair (modify, forge, injection, reply and/or delay, etc.) users' genuine data contents intended to report.

Besides the above attacks, we also consider that some smart meters could be malfunctioning and in failure status due to meter wastage, physical malfunction, etc., which will fail to report data.

Note that we focus on preventing the external attacks from divulging user's privacy and breaking the data integrity of communication packages. Other attacks, for example, the internal attack, are beyond the scope of this study.

8.2.3 Design Goal

Considering the aforementioned system model and attack model, our design goal is to propose a lightweight data aggregation scheme achieving privacy preservation and data integrity with differential privacy and fault tolerance. Specifically, the following design goals should be achieved:

1. *Privacy Preservation:* Firstly, an external attacker $\mathscr{A}$ cannot disclose user's privacy even though $\mathscr{A}$ can eavesdrop the communication flows. Secondly, $\mathscr{A}$ cannot launch differential attack to obtain the individual user's private usage data successfully. Finally, although $\mathscr{A}$ can deploy some undetectable malwares to GW or CC, it still cannot disclose user's private usage data.
2. *Authentication and Data Integrity:* A user's communication package should be authenticated that it is really transmitted by the corresponding legal residential user. The valid communications cannot be modified during the transmission, i.e., if $\mathscr{A}$ forged, altered, and/or replayed a report, the malicious behaviors should be detected.
3. *Fault Tolerance:* The system can still aggregate the data of functioning meters effectively and efficiently even in presence of malfunctioning ones.
4. *Computation Efficiency:* The computation efficiency should be achieved in the proposed protocol to support thousands and millions of residential users' data aggregation.

8.3 Preliminary

In this section, we briefly recall some preliminaries for the construction of our secure and lightweight differentially private data aggregation scheme with fault tolerance.

8.3.1 Differential Privacy

Differential privacy was first proposed by Dwork in [13] in 2006. By adding appropriately chosen noises, e.g., from Symmetric Geometric distribution, Laplace distribution, etc., to the aggregation results, the outputs will become indistinguishable with similar inputs (data sets). We call a randomized algorithm A satisfies ε-differentially privacy, if for any two data sets D_1 and D_2 differing on a single element, for all $S \subset \text{Range}(A)$, $Pr[A(D_1) \in S] \leq \exp(\varepsilon) \cdot Pr[A(D_2) \in S]$ holds.

8.3.2 *Differential Privacy via Symmetric Geometric Noise*

The use of geometric distribution to generate the noise was first put forward by Ghosh et al. in [15]. Specifically, the noise is chosen from the Symmetric Geometric distribution Geom(α), for $0 < \alpha < 1$, which can be viewed as a discrete approximation of Laplace distribution $Lap(\lambda)$ (where $\alpha \approx \exp(-\frac{1}{\lambda})$). The probability density function (PDF) of the geometric distribution Geom(α) is $Pr[X = x] = \frac{1-\alpha}{1+\alpha}\alpha^{|x|}$. Formally, if the sensitivity of the aggregation function $A(D)$ is $\Delta A = \max_{D_1,D_2} ||A(D_1) - A(D_2)||_1$ for all the data sets D_1 and D_2 differing in at most one element, then by adding geometric noise r randomly chosen from $\text{Geom}(\exp(-\frac{\varepsilon}{\Delta A}))$ to the original aggregation, the perturbed results can achieve ε-differential privacy, i.e., for any integer $k \in Range(A)$, $Pr[A(D_1) + r = k] \leq \exp(\varepsilon) \cdot Pr[A(D_2) + r = k]$ holds.

8.4 Our Proposed Scheme

In this section, we propose our lightweight data aggregation scheme achieving privacy preservation and data integrity with differential privacy and fault tolerance for smart grid communications. In order to support differential privacy, the encrypted aggregations are perturbed by GW to protect user's privacy against the differential attack. Specifically, because $g_\gamma = H_2(t_\gamma)$ can be always computed by GW for the current time point t_γ, where H_2 is the system-wide public hash function which will be defined in the following *system initialization* Section, after randomly choosing a noise from the geometric distribution, GW could perturb the aggregation by simply multiplying the encrypted noise. In order to achieve ε-differential privacy, for a given ε, according to the sensitivity of aggregation, GW first carefully chooses the parameters of geometric distribution, then computes and adds the noise, which is randomly chosen from the geometric distribution, to the original aggregation to generate the noisy counterparts.

8.4.1 *System Initialization*

The single trusted authority (TA) is in charge of bootstrapping the whole system in the beginning. Specifically, in the system initialization phase, TA executes the following steps:

- Given the security parameters τ, TA runs the parameter generator algorithm $\zeta(\tau)$ to obtain the tuple $(\mathbb{G}, p, h)$, where $\mathbb{G}$ is a cyclic group of prime order p, in which the discrete logarithm problem (DLP) is hard, and $h \in \mathbb{G}$ is a random generator of $\mathbb{G}$.

- TA defines two different public cryptographic hash functions $H_1 : \{0, 1\}^* \rightarrow \mathbb{G}$ and $H_2 : \{0, 1\}^* \rightarrow \mathbb{G}$.
- TA performs the following steps to assign key materials to residential users $U = \{U_1, U_2, \ldots, U_n\}$, GW and CC:
  - For each user $U_i \in U$, with the identity ID_i, TA chooses a random number $s_i \in \mathbb{Z}_p^*$, assigns it as $U_i's$ private key, and computes $S_i = h^{s_i}$ as $U_i's$ public key.
  - For CC, with the identity ID_c, TA computes $s_0 \in \mathbb{Z}_p^*$ satisfying $\sum_{i=0}^{n} s_i = 0 \bmod p$, assigns s_0 as CC′s private key, and computes $S_0 = h^{s_0}$ as CC′s public key.
  - For GW, with the identity ID_g, TA selects a random number $s_g \in \mathbb{Z}_p^*$, assigns it as GW′s private key, and computes $S_g = h^{s_g}$ as GW′s public key.
- TA publishes $< \mathbb{G}, p, h, H_1, H_2, S_g, S_0, ID_g, ID_c >$ and each $< S_i, ID_i >$, for $U_i \in U$, as the system-wide public information.

Meanwhile, the advanced encryption standard (AES) [23] is adopted in our system as the symmetric encryption. Denote AES-ENC$_k$ and AES-DEC$_k$ as the encryption and decryption algorithms under the symmetric key k, respectively.

8.4.2 Data Aggregation Request

Assume the reporting time points of our system, e.g., every 15 min, are defined as $T = \{t_1, t_2, \ldots, t_{max}\}$ for a sufficient long runtime period. At each time point t_γ, CC launches a *request* for collecting the usage data in the RA as follows:

Step 1: CC computes the hash value $h_\gamma = H_1(t_\gamma)$.
Step 2: CC selects a random $r \in \mathbb{Z}_p^*$, and computes $A = h_\gamma^r$.
Step 3: The computed value A is sent to GW by CC.

8.4.3 Data Aggregation Request Relay

After receiving A from CC, GW relays the data aggregation request by sending $A = h_\gamma^r$ to each $U_i \in U$ in RA, respectively.

8.4.4 User Report Generation

Recalling the stipulation in *system model* of Sect. 8.2.1 that any residential user $U_i \in U$ cannot report its measurement individually and directly to GW, suppose $U_i \in \mathrm{CL}_i$, and U_{h_i} is the CH of CL$_i$, then U_i and U_{h_i} perform the following steps collaboratively to report the measurements:

Step 1: Each user $U_i \in CL_i$ forwards its measurement to U_{h_i} as follows:

- U_i collects its usage data $m_i \in \{0, 1, \ldots, W\}$ at time point t_γ.
- U_i computes the hash values $g_\gamma = H_2(t_\gamma)$ and $h_\gamma = H_1(t_\gamma)$ for the current reporting time point t_γ.
- U_i encrypts its usage data m_i, using private key s_i, as $C_i = g_\gamma^{m_i} h_\gamma^{s_i}$ and $\delta_i = A^{s_i} = h_\gamma^{rs_i}$.
- U_i computes the noninteractive session key shared with U_{h_i}, the CH of CL_i, as $k_{ih_i} = H_1(S_{h_i}^{s_i}||ID_i||ID_{h_i}||g_\gamma) = H_1(h^{s_{h_i}s_i}||ID_i||ID_{h_i}||g_\gamma)$ and performs AES encryption using k_{ih_i} as $C_i' = \text{AES-ENC}_{k_{ih_i}}(C_i||\delta_i||ID_i||ID_{h_i}||g_\gamma)$.
- U_i sends $< C_i', ID_i >$ to U_{h_i}.

Step 2: After receiving all the messages from each user $U_i \in CL_i$, U_{h_i} pre-aggregates and reports all users' measurements within CL_i to GW as follows:

- For each received $< C_i', ID_i >$, according to ID_i, U_{h_i} computes the corresponding noninteractive session key shared with U_i as $k_{h_ii} = H_1(S_i^{s_{h_i}}||ID_i||ID_{h_i}||H_2(t_\gamma)) = H_1(h^{s_is_{h_i}}||ID_i||ID_{h_i}||g_\gamma) = k_{ih_i}$, and decrypts each received C_i' as $\text{AES-DEC}_{k_{h_ii}}(C_i') = C_i||\delta_i||ID_i||ID_{h_i}||g_\gamma$.
- Similar as CM $U_i \in CL_i$, CH U_{h_i} encrypts its own usage data m_{h_i}, as $C_{h_i} = g_\gamma^{m_{h_i}} h_\gamma^{s_{h_i}}$ and $\delta_{h_i} = A^{s_{h_i}} = h_\gamma^{rs_{h_i}}$.
- U_{h_i} pre-aggregates all the encrypted measurements of $U_i \in CL_i$ including that of itself as $C_{U_i \in CL_i} = \prod_{U_i \in CL_i} C_i = g_\gamma^{\sum_{U_i \in CL_i} m_i} h_\gamma^{\sum_{U_i \in CL_i} s_i}$ and $\delta_{U_i \in CL_i} = \prod_{U_i \in CL_i} \delta_i = (h_\gamma^{\sum_{U_i \in CL_i} s_i})^r$
- U_{h_i} computes the noninteractive session key shared with the GW as $k_{h_ig} = H_1(S_g^{s_{h_i}}||ID_{h_i}||ID_g||g_\gamma) = H_1(h^{s_gs_{h_i}}||ID_{h_i}||ID_g||g_\gamma)$, and performs AES encryption using k_{h_ig} as $C'_{U_i \in CL_i} = \text{AES-ENC}_{k_{h_ig}}(C_{U_i \in CL_i}||\delta_{U_i \in CL_i}||ID_{h_i}||ID_g||g_\gamma) = \text{AES-ENC}_{k_{h_ig}}(g_\gamma^{\sum_{U_i \in CL_i} m_i} h_\gamma^{\sum_{U_i \in CL_i} s_i}|| (h_\gamma^{\sum_{U_i \in CL_i} s_i})^r|| ID_{h_i} ||ID_g||g_\gamma)$.
- U_{h_i} reports $< C'_{U_i \in CL_i}, ID_{h_i} >$ to GW.

8.4.5 Secure Report Aggregation

Let D be any subset of residential users, and $A(D) = \sum_{U_i \in D} m_i$, then for any two data sets D_1 and D_2 differing in at most one element, $|A(D_1) - A(D_2)| \leq W$ holds, therefore, the sensitivity of A is $\Delta A = W$.

If all n smart meters work correctly, after receiving total w number of CH's reports, i.e., $< C'_{U_i \in CL_i}, ID_{h_i} >$, for $i = 1, \cdots, w$, at time point t_γ, GW executes the following steps to aggregate the measurements of all residential users $U_i \in U$:

Step 1: For each received < $C'_{U_i \in \mathrm{CL}_i}, ID_{h_i}$ >, according to ID_{h_i}, GW computes the corresponding noninteractive session key shared with U_{h_i} as $k_{gh_i} = H_1(S_{h_i}^{s_g}||ID_{h_i}||ID_g||H_2(t_\gamma)) = H_1(h^{s_{h_i}s_g}||ID_{h_i}||ID_g||g_\gamma) = k_{h_ig}$, and decrypts each $C'_{U_i \in \mathrm{CL}_i}$ as AES-DEC$_{k_{gh_i}}$ $C'_{U_i \in \mathrm{CL}_i} = g_\gamma^{\sum_{U_i \in \mathrm{CL}_i} m_i} h_\gamma^{\sum_{U_i \in \mathrm{CL}_i} s_i} ||(h_\gamma^{\sum_{U_i \in \mathrm{CL}_i} s_i})^r||ID_{h_i}||ID_g||g_\gamma = C_{U_i \in \mathrm{CL}_i}||\delta_{U_i \in \mathrm{CL}_i}||ID_{h_i}||ID_g||g_\gamma$.

Step 2: GW aggregates all the received ciphertexts (i.e., multiplying all $C_{U_i \in \mathrm{CL}_i}$, for $i = 1, \cdots, w$), such that the encrypted aggregations of all residential users $U_i \in U$ can be obtained as $C_\gamma = \prod_{i=1}^{w} C_{U_i \in \mathrm{CL}_i} = \prod_{i=1}^{w} g_\gamma^{\sum_{U_i \in \mathrm{CL}_i} m_i} h_\gamma^{\sum_{U_i \in \mathrm{CL}_i} s_i} = \prod_{U_i \in U} C_i = g_\gamma^{\sum_{i=1}^{n} m_i} h_\gamma^{\sum_{i=1}^{n} s_i}$ actually.

Step 3: GW computes the sensitivity of the aggregation as $\Delta A = W$, randomly chooses a noise $\tilde{m}$ from the geometric distribution Geom(exp($-\frac{\varepsilon}{W}$)), and computes the final aggregation as $\widetilde{C_\gamma} = C_\gamma \cdot g_\gamma^{\tilde{m}}$.

Step 4: GW computes the noninteractive session key shared with CC as $k_{gc} = H_1(S_0^{s_g}||ID_g||ID_c||g_\gamma) = H_1(h^{s_0 s_g}||ID_g||ID_c||g_\gamma)$, and performs AES encryption using k_{gc} as $\widetilde{C_g} =$ AES-ENC$_{k_{gc}}(\widetilde{C_\gamma}||ID_g||ID_c||g_\gamma) =$ AES-ENC$_{k_{gc}}(g_\gamma^{\sum_{i=1}^{n} m_i + \tilde{m}} h_\gamma^{\sum_{i=1}^{n} s_i}||ID_g||ID_c||g_\gamma)$.

Step 5: GW reports $\widetilde{C_g}$ to CC for further computation.

Note that, if smart meters of some users $\hat{U} \subset U$ do not work, i.e., $\hat{U}$ will not report their data at time point t_γ, then GW performs the following steps for privacy-preserving secure report aggregation:

Step 1: For each received < $C'_{U_i \in \mathrm{CL}_i}, ID_{h_i}$ >, according to ID_{h_i}, GW computes the corresponding noninteractive session key shared with U_{h_i} as $k_{gh_i} = H_1(S_{h_i}^{s_g}||ID_{h_i}||ID_g||H_2(t_\gamma)) = H_1(h^{s_{h_i}s_g}||ID_{h_i}||ID_g||g_\gamma) = k_{h_ig}$, and decrypts each $C'_{U_i \in \mathrm{CL}_i}$ as AES-DEC$_{k_{gh_i}}$ $C'_{U_i \in \mathrm{CL}_i} = C_{U_i \in \mathrm{CL}_i}||\delta_{U_i \in \mathrm{CL}_i}||ID_{h_i}|| ID_g ||g_\gamma$.

Step 2: GW aggregates all the received ciphertexts similar as the corresponding procedures when all the smart meters work correctly (i.e., multiplying all $C_{U_i \in \mathrm{CL}_i}$ parts), such that the encrypted aggregations of functioning users $U_i \in U/\hat{U}$ can be obtained as $\widehat{C_\gamma} = \prod_{U_i \in U/\hat{U}} C_i = g_\gamma^{\sum_{U_i \in U/\hat{U}} m_i} h_\gamma^{\sum_{U_i \in U/\hat{U}} s_i}$ actually.

Step 3: GW computes the sensitivity of the aggregation as $\Delta A = W$, randomly chooses a noise $\tilde{m}$ from the geometric distribution Geom(exp($-\frac{\varepsilon}{W}$)), and computes the final aggregation as $\widetilde{\widehat{C_\gamma}} = \widehat{C_\gamma} \cdot g_\gamma^{\tilde{m}}$.

Step 4: Similar as *Step 2*, GW also aggregates the additional auxiliary ciphertext of functioning users $U_i \in U/\hat{U}$ as $\delta_\gamma = \prod_{U_i \in U/\hat{U}} \delta_i = (h_\gamma^{\sum_{U_i \in U/\hat{U}} s_i})^r$.

Step 5: GW computes the noninteractive session key shared with CC as $k_{gc} = H_1(S_0^{s_g}||ID_g||ID_c||g_\gamma) = H_1(h^{s_0 s_g}||ID_g||ID_c||g_\gamma)$, and performs AES encryption using k_{gc} as $\widetilde{\widehat{C_g}} =$ AES-ENC$_{k_{gc}}(\widetilde{\widehat{C_\gamma}}||\delta_\gamma||ID_g||ID_c||g_\gamma) =$ AES-ENC$_{k_{gc}}(g_\gamma^{\sum_{U_i \in U/\hat{U}} m_i + \tilde{m}} h_\gamma^{\sum_{U_i \in U/\hat{U}} s_i}||(h_\gamma^{\sum_{U_i \in U/\hat{U}} s_i})^r|| ID_g||ID_c||g_\gamma)$.

Step 6: GW reports $\widetilde{\widehat{C_g}}$ to CC when malfunctioning smart meters occur.

8.4.6 Secure Report Reading

If all n smart meters work correctly, after receiving $\widetilde{C}_g$ from GW, at time point t_γ, CC performs the following steps to obtain the aggregation with Geometric noise, which achieves ε-differential privacy:

Step 1: CC computes the noninteractive session key shared with GW as $k_{cg} = H_1(S_g^{s_0}||ID_g||ID_c||H_2(g_\gamma)) = H_1(h^{s_g s_0}||ID_g||ID_c||g_\gamma) = k_{gc}$, and decrypts the received $\widetilde{C}_g$ as AES-DEC$_{k_{cg}}(\widetilde{C}_g) = g_\gamma^{\sum_{i=1}^n m_i+\tilde{m}} h_\gamma^{\sum_{i=1}^n s_i}||ID_g||ID_c||$ $g_\gamma = \widetilde{C}_\gamma||ID_g||ID_c||g_\gamma$.

Step 2: CC computes $\widetilde{C}_\gamma h_\gamma^{s_0} = g_\gamma^{\sum_{i=1}^n m_i+\tilde{m}} h_\gamma^{\sum_{i=0}^n s_i} = g_\gamma^{\sum_{i=1}^n m_i+\tilde{m}}$.

Step 3: Since $m_i \in \{1, 2, \ldots, W\}$, we have $\sum_{i=1}^n m_i \leq nW$. By computing the discrete log of $g_\gamma^{\sum_{i=1}^n m_i+\tilde{m}}$, CC can get the noisy aggregation of users' measurements as $M_{sum} = \sum_{i=1}^n m_i + \tilde{m}$ in expected time $O(\sqrt{nW})$ using Pollard's lambda method [24].

Note that, if smart meters of some users $\hat{U} \subset U$ do not work, $\widetilde{\widetilde{C}}_\gamma$ and δ_γ can be obtained at CC by computing k_{cg}, and performing AES decryption as AES-DEC$_{k_{cg}}(\widetilde{\widetilde{C}}_g) = g_\gamma^{\sum_{U_i \in U/\hat{U}} m_i+\tilde{m}} h_\gamma^{\sum_{U_i \in U/\hat{U}} s_i}||(h_\gamma^{\sum_{U_i \in U/\hat{U}} s_i})^r|| ID_g|| ID_c||g_\gamma =$ $\widetilde{\widetilde{C}}_\gamma||\delta_\gamma||ID_g||ID_c||g_\gamma$. Then CC performs the following procedures to recover the noisy aggregation of the functioning smart meters:

Step 1: CC computes $\widetilde{\widetilde{C}}_\gamma/\delta_\gamma^{\frac{1}{r}} = g_\gamma^{\sum_{U_i \in U/\hat{U}} m_i+\tilde{m}} h_\gamma^{\sum_{U_i \in U/\hat{U}} s_i}/((h_\gamma^{\sum_{U_i \in U/\hat{U}} s_i})^r)^{\frac{1}{r}} =$ $g_\gamma^{\sum_{U_i \in U/\hat{U}} m_i+\tilde{m}}$.

Step 2: Similar as the corresponding procedures when all smart meters work correctly, the aggregation of the noisy measurements of the functioning smart meters can be recovered as $\sum_{U_i \in U/\hat{U}} m_i + \tilde{m}$ successfully.

8.5 Security Analysis

In this section, we will illustrate that our proposed data aggregation scheme achieves all the security requirements defined in Sect. 8.2.

- *The user's electricity usage privacy is protected from eavesdropping.*
  Firstly, an adversary $\mathscr{A}$ may reside in RA to eavesdrop the communication flows. Suppose $\mathscr{A}$ has eavesdropped the report from U_i to CH at time point t_γ as $< C'_i, ID_i >$. Because user's measurement is encrypted in the AES ciphertext C'_i = AES-ENC$_{k_{ih_i}}(C_i||\delta_i||ID_i||ID_{h_i}||g_\gamma)$, $\mathscr{A}$ cannot obtain the corresponding plaintext, provided that the session key k_{ih_i} for AES encryption, which is generated cooperatively by utilizing the mutual communication parties' private keys, is secure against $\mathscr{A}$. In the following, we will show that even though the session key k_{ih_i} for some time point, is exposed to $\mathscr{A}$, who still cannot

obtain user's specific measurement from AES plaintexts of $C_i = g_\gamma^{m_i} h_\gamma^{s_i}$ and $\delta_i = A^{s_i} = h_\gamma^{rs_i}$. Observing the electricity usage m_i within 15 min is commonly a small value, $\mathscr{A}$ may try to launch the brute-force attack by exhaustedly testing each possible value of m_i. However, due to the discrete logarithm problem (DLP), the adversary $\mathscr{A}$ is not able to obtain $u_i's$ private key s_i, and cannot recover $U_i's$ usage data m_i. Similarly, the communications from CH to GW, and from GW to CC are of the same form as U_i's report to CH, thus, $\mathscr{A}$ cannot obtain the individual user's usage data via eavesdropping the communication flows.

When some smart meters, say $\hat{U} \subset U$, are malfunctioning, because the value of r is kept secret by CC, anyone else cannot use r to recover the sum usage of functioning smart meter as $\sum_{U_i \in U/\hat{U}} m_i + \tilde{m}$, let alone each user's private usage data m_i, even if the session key k_{gc} for AES encryption between GW and CC is exposed to $\mathscr{A}$ and who could obtain the corresponding plaintexts of $\widetilde{\widetilde{C_\gamma}}$ and δ_γ of some time point.

- *The user's electricity usage privacy is protected from malware attack.*
  Even though the adversary $\mathscr{A}$, after deploying some undetectable malwares into GW or intruding into the database of GW, has stolen the stored data successfully, who could only get the aggregations and ciphertexts of all users' data. Because GW never decrypts any user's electricity usage data, the adversary $\mathscr{A}$ still cannot get any user's private usage data. In addition, $\mathscr{A}$ could also intrude into the database of CC, however, after decryption, the outputs that CC generated are all noisy aggregations of users' data, which do not reveal individual user's usage data at all. Therefore, the individual user's report is protected from malwares attack.
- *The user's electricity usage privacy is protected from differential attack.*
  For a given privacy level ε, GW perturbs the aggregations without decryption by adding appropriate geometric noises in the form of ciphertext. By this means, ε-differential privacy is achieved. Specifically, GW adds the noise $\tilde{m}$, which is chosen from $\mathrm{Geom}(\exp(-\frac{\varepsilon}{W}))$, to the exact aggregation to obtain the perturbed one. Assume the adversary $\mathscr{A}$ acquires two perturbed aggregations $s + \tilde{m}^{(s)}$ and $t + \tilde{m}^{(t)}$, where s and t are two aggregations of the two data sets differing in at most one element, respectively, while $\tilde{m}^{(s)}$ and $\tilde{m}^{(t)}$ are the two corresponding geometric noises. Similar to the deduction in [10], since $|s - t| \leq W$, for any integer k, we have $\mu = Pr[s + \tilde{m}^{(s)} = k]/Pr[t + \tilde{m}^{(t)} = k] = Pr[\tilde{m}^{(s)} = k - s]/Pr[\tilde{m}^{(t)} = k - t] = (\frac{1-\alpha}{1+\alpha}\alpha^{|k-s|})/(\frac{1-\alpha}{1+\alpha}\alpha^{|k-t|}) = \alpha^{|k-s|-|k-t|}$. Since $-|s - t| \leq |k - s| - |k - t| \leq |s - t|$ and $0 < \alpha < 1$, we have $e^{-\varepsilon} = (e^{-\frac{\varepsilon}{W}})^W \approx \alpha^W \leq \alpha^{|s-t|} \leq \mu \leq \alpha^{-|s-t|} \leq \alpha^{-W} \approx (e^{-\frac{\varepsilon}{W}})^{-W} = e^{\varepsilon}$. Thus, ε-differential privacy is satisfied. Therefore, even though the adversary $\mathscr{A}$ obtains the aggregations of two similar data sets, via launching differential attacks, the individual user's privacy is still not leaked at all.
- *The users' electricity usage data aggregation is secure and reliable with the function of fault tolerance of report failures.*
  We innovate a new distributed method to realize fault tolerance of users' report failures. Even under the circumstances when $\hat{U} \subset U$ do not work, the value r,

which is kept secret by CC, together with the aggregated auxiliary ciphertext δ_γ still can be used to recover the aggregated data of $\sum_{U_i \in U/\hat{U}} m_i + \hat{m}$. Specifically, CC, after decrypting and obtaining $\widetilde{\widetilde{C_\gamma}}$ and δ_γ, uses the private information of r to recover the sum usage of functioning smart meters. Because r is kept secret by CC, without it, anyone else cannot recover the sum of functioning smart meters' usage data, not to mention each user's private usage data m_i.

- *The user's communication link is protected from data alteration attack.*

  We will show that user's report can be authenticated that it is really sent by a legal residential user, and the communications cannot be altered during the transmissions.

  - Source Authentication

    Firstly, we consider the communications from CM U_i to CH. U_i first generates the noninteractive session key shared with CH as $k_{ih_i} = H_1(S_{h_i}^{s_i} || ID_i || ID_{h_i} || H_2(t_\gamma))$ using U_i's secret key s_i and CH's public key S_{h_i}. Then the report is encrypted using the session key k_{ih_i}, and the generated ciphertext together with U_i's identity ID_i are transmitted to CH. After receiving the report, according to ID_i, CH computes the same noninteractive session key shared with U_i as $k_{h_i i} = H_1(S_i^{s_{h_i}} || ID_i || ID_{h_i} || H_2(t_\gamma)) = k_{ih_i}$, then uses $k_{h_i i}$ to decrypt the report properly. It is obvious that only if the report came from the legal CM, can it be decrypted correctly, thereby, the source authentication can be ensured. Due to the same reason, the source authentication of the report from CH to GW and from GW to CC can be ensured similarly. In summary, the proposed scheme achieves source authentication in the whole communications.

  - Data Integrity

    - Communications Pollution Attack Resistance

      Firstly, we consider the communications from U_i to CH. Upon receiving $< C_i', ID_i >$, according to ID_i, CH first computes the noninteractive session key $k_{h_i i} = H_1(S_i^{s_{h_i}} || ID_i || ID_{h_i} || H_2(t_\gamma)) = H_1(h^{s_i s_{h_i}} || ID_i || ID_{h_i} || g_\gamma) = k_{ih_i}$ shared with U_i. Then, CH performs the AES decryption using $k_{h_i i}$ to obtain U_i's report. Because the secret keys s_i and s_{h_i}, which are kept secret by CM U_i and CH U_{h_i}, respectively, are utilized to compute the shared session key $k_{h_i i}$ (k_{ih_i}) collaboratively, the external adversary $\mathscr{A}$ cannot obtain the agreed secret key, nor can it alter the original data encrypted and reported by U_i. Because a pair of identities of two communication parties are incorporated into the one-way hash function to generate the noninteractive session key and AES encryption scheme is secure, even the insider legal participants cannot forge a new valid report to impersonate and frame the innocent residential user. Therefore, the communications from CM to CH cannot be polluted maliciously. Due to the same reason, the data integrity of the report from CH to GW and from GW to CC can be ensured similarly. In summary, the proposed scheme achieves data integrity throughout the whole communications.

- Message Replay Attack Resistance

  In the proposed scheme, after receiving the message $< C_i', ID_i >$ from U_i, CH decrypts C_i' to obtain $C_i||\delta_i||ID_i||ID_{h_i}||g_\gamma$, then according to the current time point t_γ, computes $H_2(t_\gamma)$ and checks whether $g_\gamma = H_2(t_\gamma)$ holds for each CM. Because only the fresh report corresponding to current time point t_γ can pass the verification, the proposed scheme can resist the message replay attack.

8.6 Performance Evaluation

The proposed scheme achieves privacy preservation and data integrity simultaneously for secure data aggregation with differential privacy and fault tolerance for smart grid communications. In this section, we will mainly compare the performance of our proposed scheme with the state-of-the-art similar schemes [6, 11, 19]. The major functions and features of these systems are first compared in Table 8.1. Both [11] and [19] support data integrity for smart grid communications. Meanwhile, privacy preservation is supported by the schemes of [6] and [11]. Because all users' blinding factors should be utilized cooperatively to obtain the sum electricity usage in [6] and [11], they cannot support fault tolerance of malfunctioning smart meters. Observing that the regular peer-to-peer communication architecture is used in [19], even though theoretically it could be trivially applied in the circumstances when malfunctioning smart meters occur, comparisons will show that our proposed scheme achieves remarkable advantages in terms of efficiency. Because most of the computations of the *hub-like* entity GW are decentralized to CHs in RA, our proposed scheme is very efficient and scalable to support thousands even millions of residential users' data aggregation. Thus, our comparison is focused on data report in user side, which includes both computation cost and communication overhead. In addition, different from all the aforementioned similar works [6, 11, 19], differential privacy and fault tolerance are taken into consideration at the same time in our scheme. As a result, we also compare our proposed scheme with the state-of-the-art data aggregation schemes [5], which supports differential privacy and fault tolerance, in terms of utility of differential privacy and robustness of fault tolerance.

- *Comparison of Computation Complexity*

  For our proposed scheme, when a CM U_i reports the measurement, it requires one multiplication, three hash, and four exponentiation operations to compute the ciphertext of usage data and non-interactive session key shared with CH, in which one exponentiation operation can be pre-computed beforehand in off-line phase. And one AES encryption operation is cost to encrypt the report as well. Suppose the size of the cluster CL_i is n_i, after receiving the ciphertexts from (n_i-1) CMs, CH first computes all non-interactive session keys shared with each CM within the same cluster CL_i, and decrypts the received ciphertext. It will cost

Table 8.1 Feature comparison

	Proposed scheme	Fan et al.'s scheme [11]	Fouda et al.'s scheme [19]	Erkin and Tsudik's scheme [6]
D	Yes	Yes	Yes	No
P	Yes	Yes	No	Yes
F	Yes	No	Partial[a]	No

D: data integrity, P: privacy preservation, F: supporting data aggregation with fault tolerance
[a]Because the simplex and generic peer-to-peer communication architecture is considered, it cannot be regarded as having achieved fault tolerance completely

n_i-1 exponentiation, n_i+1 hash, and n_i-1 AES decryption operations, in which n_i-1 exponentiation operations can be pre-computed beforehand in off-line phase. Subsequently, 3 exponentiation, 2n_i-1 multiplication operations are needed to encrypt CH's usage data and aggregate with the other n_i-1 CM's reports. Finally, 1 AES encryption operation is cost to achieve data integrity. After receiving all the reports from w CHs, GW first computes the noninteractive session key shared with each CH, and decrypts each report. It will cost w+1 hash, w exponentiation, and w AES decryption operations, in which w exponentiation can be pre-computed beforehand in off-line phase. Subsequently, 2(w-1) multiplication operations are cost to aggregate all the received residential users' ciphertexts. Finally, 1 AES encryption operation is cost to achieve data integrity. Because the entity CC, who is in charge of monitoring and controlling the whole smart gird, is introduced in our system model, and it is not considered in the other three schemes [6, 11, 19], the corresponding comparisons are not performed here.

In the scheme of [19], each user reports the data to the aggregator in peer-to-peer way. Firstly, one public key encryption, one hash operations, one public key decryption, and two exponentiation are needed for each user to agree on the session key with the aggregator. Then, one HMAC generation, and one AES encryption operations are cost for each user to transmit the authenticated report to the aggregator. Subsequently, it will also cost one public key encryption, one hash, one public key decryption, and two exponentiation operations for the aggregator to agree on the session key with each user, and cost one AES decryption, and one HMAC verification operations for the aggregator to verify the integrity of each user's report. For n users, n sets of such operations are needed.

In the scheme of [11], firstly, for each residential user, it costs three hash, two exponentiation, one addition, and one multiplication operations to generate the registration information. Then, one hash, two multiplication, and three exponentiation operations are needed to encrypt the measurement, and one exponentiation and one hash operations are cost to sign on the measurement. Subsequently, n+1 pairing, 3n-2 multiplications, 2n+2 exponentiation, n+1 hash, and 1 2-DNF formulas cryptosystem decryption operations are needed for the aggregator to aggregate and recover the sum usage data.

In the scheme of [6], firstly, for each residential user, it costs one multiplication, $2n$ addition, and two exponentiation operations to exchange random numbers and encrypt the individual measurement, where n is the total number of users. Then, n-1 multiplication and 1 2-DNF formulas cryptosystem decryption operations are needed for the aggregator to aggregate and recover the sum usage data.

Note that one of the remarkable advantages of our proposed scheme over the other three schemes [6, 11, 19], lies in the characteristic of decentralizing the computational cost of the *hub-like* entity GW. The number of the total clusters w and the scale of each cluster n_i, for $i = 1, \cdots, w$, are configurable. Here, for quantitative comparison, without loss of generality, we set all parameters of n_i to 10 in our proposed scheme, where $i = 1, \cdots, w$, and thus w equals to $\lceil n/10 \rceil$. For clear illustration, we first abbreviate all the operation notations as enumerated in Table 8.2. Then, the computation cost of each user and the aggregator (for our proposed scheme, the corresponding computation cost includes both CH and GW) of the four schemes are compared in Table 8.3. Furthermore, we perform the experiments with MIRACL [25, 26] library and JPBC (Java Pairing Based Cryptography) library [27, 28] running on a 3.0 GHz processor Pentium IV system to study the operation cost. The time cost of all the primitive operations are indicated in Table 8.2 as well. Based on the test results, we plot the comparisons of computation cost in Fig. 8.3. It can be seen clearly from the figure that our scheme greatly reduces the computation complexity of both user side and aggregator side.

- *Comparison of Communication Cost*

  The communications of the proposed scheme consists of three parts, CM to CH, CH to GW, and GW to CC. Because via introducing the system-wide trustable entity CC, our proposed scheme achieves enhanced security characteristics which are not considered in other three schemes [6, 11, 19], here,

Table 8.2 Time cost of operations

Notations	Descriptions	Time cost
C_a	Addition	$\approx$ 0.004 ms
C_m	Multiplication	$\approx$ 0.16 ms
C_e	Exponentiation	$\approx$ 1.7 ms
C_H	Hash	$\approx$ 0.0037 ms
C_{HM}	HMAC	$\approx$138 MiB/s
$C_{\mathrm{HM_V}}$	HMAC verification	$\approx$138 MiB/s
$C_{\mathrm{AES_E}}$	AES encryption	$\approx$ 75 MiB/s
$C_{\mathrm{AES_D}}$	AES decryption	$\approx$ 75 MiB/s
$C_{\mathrm{PK_E}}$	Public key encryption	$\approx$ 0.09 ms
$C_{\mathrm{PK_D}}$	Public key decryption	$\approx$ 2.28 ms
C_p	Pairing	$\approx$ 19 ms
$C_{\mathrm{2\text{-}DNF}}$	2-DNF formulas cryptosystem decryption	$\approx$ 1.06 ms

Table 8.3 Computation cost comparisons

Protocol	User	Aggregator
Proposed scheme	$3C_H+C_m+3Ce+C_{AES_E}$	$nC_{AES_D}+(1.9n-2)C_m+(1.2n+1)C_H+0.3nC_e+(0.1n+1)C_{AES_E}$
Fouda et al.'s scheme [19]	$2C_e+C_{PK_E}+C_{PK_D}+C_H+C_{HM}+C_{AES_E}$	$n(2C_e+C_{PK_E}+C_{PK_D}+C_H+C_{AES_D}+C_{HM_V})$
Fan et al.'s scheme [11]	$6C_e+C_a+3C_m+5C_H$	$(3n-2)C_m+(n+1)C_p+(2n+2)C_e+(n+1)C_H+C_{2\text{-DNF}}$
Erkin and Tsudik's scheme [6]	$C_m+2C_e+2nC_a$	$n(C_m+2C_e+2nC_a)+(n-1)C_m+C_{2\text{-DNF}}$

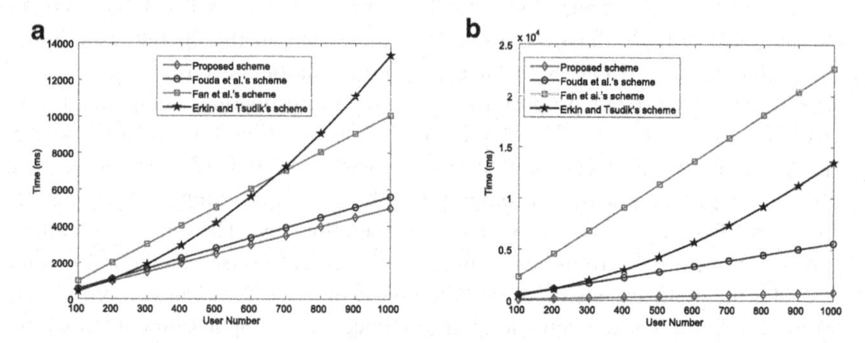

Fig. 8.3 Performance comparison of computation cost at (**a**) user side and (**b**) aggregator side

we focus on the comparison of the common parts, i.e., communication overhead between residential users and GW.

In our scheme, when CM generates and delivers the report to CH, the data report is in the form of $< C_i', ID_i >$, where $C_i' = \text{AES-ENC}_{k_{ih_i}}(C_i||\delta_i||ID_i||ID_h||g_\gamma)$ and ID_i is CM's identity. Each CM's individual communication overhead will be 1684 bits, if AES-256 encryption is chosen, and the length of ID_i and ID_{hi}, and the security parameters of τ, are set to 20-bit, 20-bit, and 512-bit, respectively. Each CH U_{h_i} collects the reports from total n_i-1 CMs, which indicates that the total communication overhead between CMs and CH is 1684(n_i-1) bits in one cluster. For all w clusters, the overall communication overhead will be $\sum_{i=1}^{w}$ 1684(n_i-1) bits totally. Subsequently, each CH aggregates and forwards the report to GW in the form of $< C'_{U_i\in CL_i}, ID_{h_i} >$, where $C'_{U_i\in CL_i} = \text{AES-ENC}_{k_{h_ig}}(C_{U_i\in CL_i}||\delta_{U_i\in CL_i}||ID_{h_i}||ID_g||g_\gamma)$, and ID_{h_i} is CH's identity. Similarly, each CH's individual communication overhead is 1684 bits, and the overall communication overhead between CHs and GW will be 1684w bits for w pairs of communications. In summary, the overall communication overhead of our proposed scheme for all n residential users to report the measurements to GW will be $\sum_{i=1}^{w} 1684(n_i\text{-}1)+1684w = 1684\sum_{i=1}^{w} n_i = 1684n$ bits.

In the scheme of [19], firstly, two public key ciphertexts and 1 q-bit element in Z_q^* are needed to be transmitted between each user and the aggregator to agree on the session key k_i. The communication overhead will be (2048+512)-bit length, if q is set to 512-bit length, and 1024-bit RSA modulus is chosen. Then, each user U_i reports the ciphertext of usage data as $\text{ENC}_{k_i}(m_i||T_i||HMAC_{k_i}(m_i||T_i))$ to the aggragator. The size will be 384 bits, if the length of $m_i||T_i$ is set to 200-bit, and AES-256 encryption and HMAC(SHA-1) are chosen. Thus, each user's total communication overhead is 2944 bits. For total n users, the overall communication overhead turns to be 2944n bits.

In the scheme of [11], firstly, $< Y_i, \alpha_i, \beta_i, r_i, ID_i >$ need to be transmitted for each user to register into the system, and the communication overhead will be 2068 bits, if 512-bit q is chosen, and the size of each user's identity is with 20-bit length. Then, each user reports $CT_i = g_0^{(H_2(t)h'_i)^{\pi_i}}$ and $\delta_i = H_1(t||CT_i)^{x_i}$ to the aggregator. The size will be 1536 bits, if 1024-bit n is chosen. The aggregator collects the reports from total n users, which indicates that the overall communication overhead between the users and the aggregator is 3604n bits.

In the scheme of [6], users should exchange random numbers and broadcast reports each other in every time point. If 1024-bit modular size is chosen, then the overall communication overhead for all n residential users to repot and aggregate the measurements will be $3n$(n-1) · 1024.

The communication overhead in terms of user number n of the four schemes are plotted in Fig. 8.4. It can be seen from the figure that our proposed scheme achieves lower communication overhead compared with the other three schemes.

From the above analysis, our proposed scheme is actually efficient in terms of computation complexity and communication cost, which is applicable for the real-time high-frequent data aggregation in smart grid communications.

- *Comparison of Robustness of Fault Tolerance*

  Because our scheme inherits all the basic functionality and performance requirements of fault tolerant smart grid data aggregation schemes, here, we focus on the comparison of our scheme with the state-of-the-art data aggregation scheme [5] in terms of robustness of fault tolerance. In the scheme of [5], GW stores B pieces of *future ciphertexts* for each smart meter to support fault tolerance. Without loss of generality, assume the data report interval of the smart grid communications of [5] is T. And suppose at time point T_a, due to some failures, certain smart meter U_i cannot report the measurement successfully to GW, and the fault recovery time point is T_b. Thus, the fault persistence period T_{per} is T_b-T_a. If T_{per} >B·T, until the fault smart meter U_i is to be recovered again on T_b, the system of [5] cannot tolerate the fault any longer after the time point of T_a+B·T, because the pre-stored *future ciphertexts* are used up. The robustness of fault tolerance turns to be much worse when the number of the malfunctioning smart meters increases. In order to support more robust fault tolerance, the system parameter of the buffer size B of [5] should be increased further. However, this causes heavy storage cost, computation complexity and communication overhead. By contrast, our scheme is more robust of fault

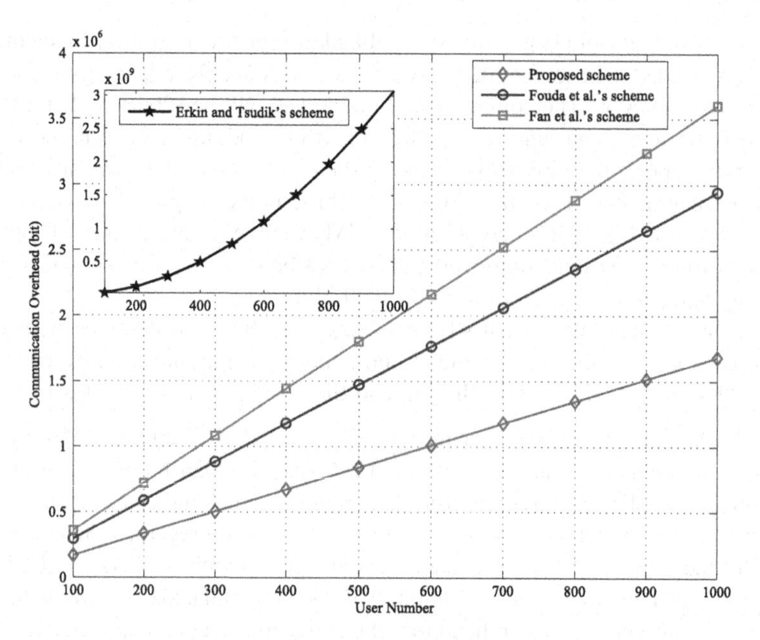

Fig. 8.4 Performance comparison of communication overhead

tolerance and can support efficient data aggregation with any rational number of malfunctioning smart meters with arbitrary long fault period, because our fault tolerance mechanism is not related to the malfunctioning smart meters directly and is independent of any external factors, e.g., *future ciphertexts*.

- *Comparison of Utility of Differential Privacy*

  Through the following comparison, we will show that our proposed data aggregation scheme provides higher utility (i.e., low error) in terms of differential privacy than the state-of-the-art data aggregation scheme of [5]. Similar to the scheme of [5], we implement an electricity trace simulator which is able to generate realistic 1-min consumption traces. We produce traces for 2000 households based on this simulator. In the scheme of [5], in order to resit the attack of substracting *current ciphertext* and *future ciphertext*, an additional Laplatics noise $Lap(\lambda)$ is added to each smart meter's *future ciphertext*. However, as commented by the authors of [5], this incurs large errors which increases greatly with the increasing number of malfunctioning smart meters. Our scheme overcomes this drawback, thus, as compared in Fig. 8.5, it is of better utility than the scheme of [5]. The figures illustrate the traces of the actual total measurements, the noisy counterparts of both [5] and our proposed scheme, for the different parameters, where in each of the figure, n and p denote the total number of the household, and the different ratio of malfunctioning smart meters, respectively. As it can be seen from the figures, the larger the number of p, the more accurate of our scheme comparing with the scheme of [5]. More precisely, let the 1-h root-mean-square-error (RMSE) of Jongho et al.'s protocol [5] and our

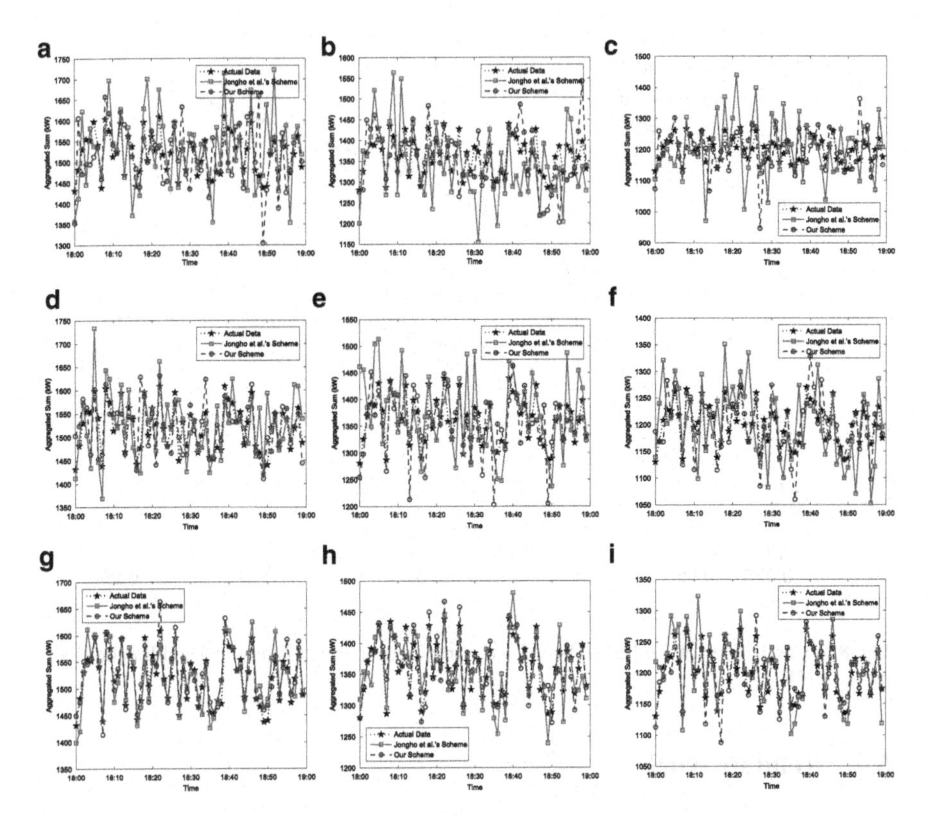

Fig. 8.5 Comparison of noisy total consumption between the proposed aggregation protocol and Jongho et al.'s protocol [5]. (**a**) n = 2000, p = 0.05, $\varepsilon = 0.5$; (**b**) n = 2000, p = 0.15, $\varepsilon = 0.5$; (**c**) n = 2000, p = 0.25, $\varepsilon = 0.5$; (**d**) n = 2000, p = 0.05, $\varepsilon = 1$; (**e**) n = 2000, p = 0.15, $\varepsilon = 1$; (**f**) n = 2000, p = 0.25, $\varepsilon = 1$; (**g**) n = 2000, p = 0.05, $\varepsilon = 2$; (**h**) n = 2000, p = 0.15, $\varepsilon = 2$; (**i**) n = 2000, p = 0.25, $\varepsilon = 2$

proposed scheme be γ_1 and γ_2, respectively. The ratios of $\gamma = \frac{\gamma_1}{\gamma_2}$ with p under different privacy level ε are depicted in Fig. 8.6, which shows that comparing with [5], our proposed scheme always achieves better utility due to much lower errors in each circumstance.

We also set ε, the differential privacy level, to 0.5, 1, 2, for various scenarios with different extent of failures. As it can be seen from Fig. 8.5, the larger ε is, the smaller noise will be added, and then the utility is higher; while the smaller ε is, the larger noise will be included, and then the higher level of the privacy can be guaranteed. Compared with the case of ε = 2, the utility in ε = 0.5 is lower, but it is still acceptable. Therefore, in real scenarios, there is a tradeoff between the privacy and utility.

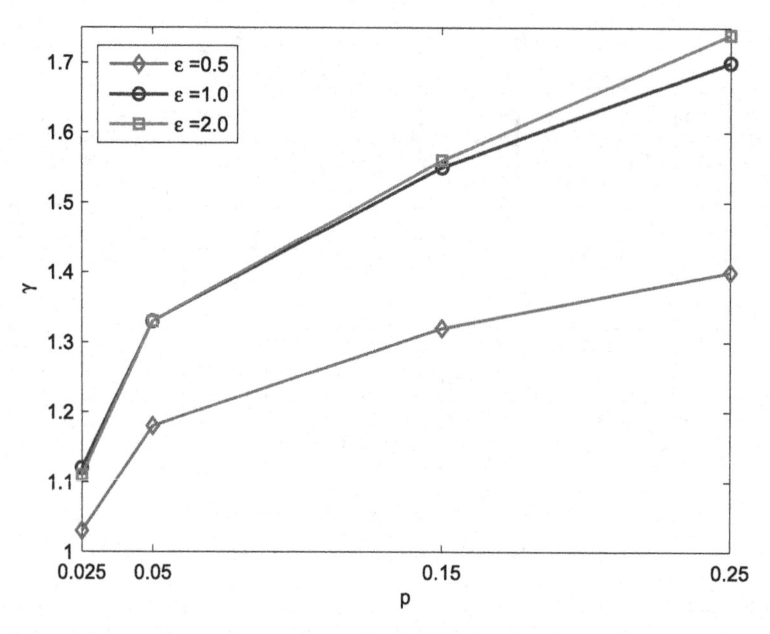

Fig. 8.6 Comparison of 1-h RMSE between the proposed aggregation protocol and Jongho et al.'s protocol [5]

8.7 Related Work

In this section, we put our emphasis on the discussion of some other literatures [5, 6, 11, 19, 20, 29, 30] related to our research which also achieve privacy preservation and/or data integrity for smart gird communications.

Exploring simple cryptographic privacy techniques, Erkin and Tsudik propose one popular privacy-preserving data aggregation scheme without any on-line aggregator or trusted third party [6]. However, users should exchange random numbers and broadcast reports each other in every time point, which incurs both individual and overall communication overhead. Li et al. propose one in-network data aggregation architecture for smart grid communications [29]. Unfortunately, the concrete scheme they designed cannot achieve data integrity. Subsequently, Li et al. present another data aggregation scheme to achieve privacy preservation and data integrity concurrently [30]. The peer-to-peer digital signature exploring homomorphic techniques is designed and the checksum of the aggregation is calculated and updated following in-network aggregation flows. However, the hop-by-hop verification process incurs huge additional storage and communication overhead and the incremental signature verifications launched by the aggregator could expose individual's privacy. Fouda et al. propose a message authentication scheme for smart grid communications [19]. Borrowing Diffie-Hellman key exchange technique, the session key is shared between each user and the gateway. Then, with the agreed session key and by employing HMAC technique, the subsequent communications can be authenticated. However, as pointed out by the authors, the security of the system is dependant on the round-based public key encryption and decryption to set up the secure session key, which leads to heavy computation and communication overhead. Besides, the scheme does not achieve privacy preservation for residential user. Alharbi et al. propose one data aggregation for smart grid communications to achieve data security and privacy preservation for residential users [20] with static topology. It is characterized by taking advantage of one-time masking technique to protect user's privacy with high efficiency. However, the session key should be shared between each user and the aggregator, and the neighbouring users also need to agree on session keys, which incurs heavy burden for key management. In [5], Jongho et al. propose a fault tolerant aggregation protocol for privacy-assured smart grid communications. The *future ciphertexts* are leveraged to support fault tolerance of possible communication failures, which leads to the heavy round-based communication, computation and storage overhead. Based on homomorphic encryption techniques, Fan et al. utilize a tree-based aggregation approach to aggregate users' reports efficiently [11]. Through distributing blind factors among all parties including each residential user and the aggregator, the scheme achieves privacy preservation. The registration procedure is interacted between the user and the aggregator to produce user's private key for generating signatures on encrypted data reports to achieve data integrity. However, the pairing based signature verification procedure is resource-consuming. In addition, after taking a close look at the interactive registration procedure, user's private key can be inferred from the public information, which sows the hazards of impairing data integrity.

Although our proposed scheme addresses the similar issues, i.e., to achieve efficient data aggregation with privacy preservation and data integrity in smart grid communications, comparing with existing works, our research emphasis still has some differences: (1) we propose our data aggregation scheme in a more challenging threaten model to resist privacy divulging attack, which covers eavesdropping attack, differential attack, and malware attack, and data alteration attack simultaneously and (2) we take enhanced properties of differential privacy and fault tolerance into consideration meanwhile; thus, it additionally improves the reliability and practicability.

8.8 Summary

In this chapter, we have proposed a secure data aggregation scheme for smart grid communications which not only achieves security properties of privacy preservation and data integrity simultaneously, but also improves the practicability and reliability due to implementation and integration of enhanced properties of differential privacy and fault tolerance. Particularly, under more challenging threaten model which covers communication attack, differential attack, and malware attack, the proposed scheme is secure against privacy divulging attack. Meanwhile, the proposed non-interactive session key agreement mechanism prevents the communications from being polluted and impaired. Through extensive performance evaluation, we have also demonstrated that the proposed scheme outperforms the state-of-the-art similar schemes in terms of computation complexity, communication cost, robustness of fault tolerance, and utility of differential privacy.

References

1. H. Bao and R. Lu, "A lightweight data aggregation scheme achieving privacy preservation and data integrity with differential privacy and fault tolerance," *Peer-to-Peer Networking and Applications*, 2015.
2. R. Lu, X. Liang, X. Li, X. Lin, and X. Shen, "Eppa: An efficient and privacy-preserving aggregation scheme for secure smart grid communications," *IEEE Transactions on Parallel and Distributed Systems*, vol. 23, no. 9, pp. 1621–1631, 2012.
3. L. Chen, R. Lu, and Z. Cao, "Pdaft: A privacy-preserving data aggregation scheme with fault tolerance for smart grid communications," *Peer-to-Peer Networking and Applications*, pp. 1–11, 2014.
4. E. Shi, T.-H. H. Chan, E. G. Rieffel, R. Chow, and D. Song, "Privacy-preserving aggregation of time-series data." in *NDSS*, vol. 2, no. 3, 2011, p. 4.
5. J. Won, C. Y. Ma, D. K. Yau, and N. S. Rao, "Proactive fault-tolerant aggregation protocol for privacy-assured smart metering," in *INFOCOM 2014*. IEEE, 2014, pp. 2804–2812.
6. Z. Erkin and G. Tsudik, "Private computation of spatial and temporal power consumption with smart meters," Springer, pp. 561–577, 2012.
7. F. D. Garcia and B. Jacobs, "Privacy-friendly energy-metering via homomorphic encryption," in *Security and Trust Management*. Springer, 2011, pp. 226–238.
8. V. Rastogi and S. Nath, "Differentially private aggregation of distributed time-series with transformation and encryption," in *Proceedings of the 2010 ACM SIGMOD International Conference on Management of data*. ACM, 2010, pp. 735–746.
9. G. Acs and C. Castelluccia, "I have a dream!(differentially private smart metering)," in *Information Hiding*. Springer, 2011, pp. 118–132.
10. L. Chen, R. Lu, Z. Cao, K. AlHarbi, and X. Lin, "Muda: Multifunctional data aggregation in privacy-preserving smart grid communications," *Peer-to-Peer Networking and Applications*, pp. 1–16, 2014.
11. C.-I. Fan, S.-Y. Huang, and Y.-L. Lai, "Privacy-enhanced data aggregation scheme against internal attackers in smart grid," *IEEE Transactions on Industrial Informatics*, vol. 10, no. 1, pp. 666–675, 2014.
12. P. Paillier, "Public-key cryptosystems based on composite degree residuosity classes," in *Advances in cryptology—EUROCRYPT'99*. Springer, 1999, pp. 223–238.
13. C. Dwork, "Differential privacy," in *Automata, languages and programming*. Springer, 2006, pp. 1–12.
14. ——, "Differential privacy: A survey of results," in *Theory and Applications of Models of Computation*. Springer, 2008, pp. 1–19.
15. A. Ghosh, T. Roughgarden, and M. Sundararajan, "Universally utility-maximizing privacy mechanisms," *SIAM Journal on Computing*, vol. 41, no. 6, pp. 1673–1693, 2012.
16. A. Perrig, "The biba one-time signature and broadcast authentication protocol," in *Proceedings of the 8th ACM conference on Computer and Communications Security*. ACM, 2001, pp. 28–37.
17. W. D. Neumann, "Horse: an extension of an r-time signature scheme with fast signing and verification," in *International Conference on Information Technology: Coding and Computing (ITCC 2004)*, vol. 1. IEEE, 2004, pp. 129–134.

18. D. Johnson, A. Menezes, and S. Vanstone, "The elliptic curve digital signature algorithm (ecdsa)," *International Journal of Information Security*, vol. 1, no. 1, pp. 36–63, 2001.
19. M. M. Fouda, Z. M. Fadlullah, N. Kato, R. Lu, and X. Shen, "A lightweight message authentication scheme for smart grid communications," *IEEE Transactions on Smart Grid*, vol. 2, no. 4, pp. 675–685, 2011.
20. K. Alharbi and X. Lin, "Lpda: a lightweight privacy-preserving data aggregation scheme for smart grid," in *2012 International Conference on Wireless Communications and Signal Processing (WCSP)*. IEEE, 2012, pp. 1–6.
21. D. A. Knox and T. Kunz, "Rf fingerprints for secure authentication in single-hop wsn," in *IEEE International Conference on Wireless and Mobile Computing, Networking and Communications, 2008. WIMOB'08*. IEEE, 2008, pp. 567–573.
22. M. Kgwadi and T. Kunz, "Securing rds broadcast messages for smart grid applications," *International Journal of Autonomous and Adaptive Communications Systems*, vol. 4, no. 4, pp. 412–426, 2011.
23. J. Daemen and V. Rijmen, *The design of Rijndael: AES-the advanced encryption standard*. Springer Science & Business Media, 2002.
24. A. J. Menezes, P. C. Van Oorschot, and S. A. Vanstone, *Handbook of applied cryptography*. CRC press, 2010.
25. M. Scott, "Miracl–multiprecision integer and rational arithmetic c/c++ library," *Shamus Software Ltd, Dublin, Ireland*, 2003.
26. P. Failla, "Privacy-preserving processing of biometric templates by homomorphic encryption," Ph.D. dissertation, Ph. D. dissertation, PhD School in Information Engineering, University of Siena, Italy, 2010.
27. M. Scott, "Implementing cryptographic pairings," *Lecture Notes in Computer Science*, vol. 4575, p. 177, 2007.
28. B. Lynn *et al.*, "Pbc: The pairing-based cryptography library," http://crypto.stanford.edu/pbc, 2011.
29. F. Li, B. Luo, and P. Liu, "Secure information aggregation for smart grids using homomorphic encryption," in *2010 First IEEE International Conference on Smart Grid Communications (SmartGridComm)*. IEEE, 2010, pp. 327–332.
30. F. Li and B. Luo, "Preserving data integrity for smart grid data aggregation," in *2012 IEEE Third International Conference on Smart Grid Communications (SmartGridComm)*. IEEE, 2012, pp. 366–371.

Printed in the United States
By Bookmasters

Printed in the United States
By Bookmasters